INTRODUCTION TO COMPUTATIONAL NEUROBIOLOGY AND CLUSTERING

Series on Advances in Mathematics for Applied Sciences – Vol. 73

INTRODUCTION TO COMPUTATIONAL NEUROBIOLOGY AND CLUSTERING

Brunello Tirozzi
Daniela Bianchi
Department of Physics
University of Rome "La Sapienza", Italy

Enrico Ferraro
Department of Biology
University of Rome "Tor Vergata", Italy

World Scientific

NEW JERSEY • LONDON • SINGAPORE • BEIJING • SHANGHAI • HONG KONG • TAIPEI • CHENNAI

Published by

World Scientific Publishing Co. Pte. Ltd.

5 Toh Tuck Link, Singapore 596224

USA office: 27 Warren Street, Suite 401-402, Hackensack, NJ 07601

UK office: 57 Shelton Street, Covent Garden, London WC2H 9HE

British Library Cataloguing-in-Publication Data
A catalogue record for this book is available from the British Library.

ISBN-13 978-981-270-539-6
ISBN-10 981-270-539-2

Printed in Singapore.

To our dear ones

Preface

The book is divided into two parts. The first concerns the simulations of neurobiological systems, the second describes the currently used algorithms of genes and protein classification. The first part of this book corresponds to the lectures that one of us is giving for the course of *Fisica Applicata alle Biotecnologie* (Applied Physics to Biotechnology). Biotechnology is a quite recent discipline taught in Italian Universities, as well as in other parts of the world, and many students moved from Biology courses to it. The aim of the book, as well as the lectures, is to teach simulations of neurobiological systems to students with a little background of physics and mathematics, a usual situation for the students of biology. In the lectures I succeeded to teach only the first part but the most curious students and readers should not stop and should go also through the second one. These two themes are very important and there is a huge amount of work on them in the current research literature. The mathematical models of neurobiology have been at the center of attention of the research in neurobiology for 20 years. Actually the story is much older since we can establish a starting point for this discipline with the model of Hodgkin and Huxley of the fifties in which the neural activity of the axon of the squid was fully described. So there is an old tradition in this field which has been renewed many times. In the last decade the models describing the neurons have become more sophisticated by the introduction of the stochastic inputs described by the Poisson and the Wiener processes. There was a transition from ordinary differential equations to stochastic differential equations and the mathematics become more complicated, stochastic contributions were inserted in the non-linear system of equations of the Hodgkin Huxley model (or of the Fitzhugh Nagumo model) as well as in the simpler Integrate & Fire model described in the first chapter of the book. Actually there is also a huge variety of non-linear models describing the neurons in different situations of the nerve system (brain, spinal chord, etc.). Each system of neurons has a very complicated structure and interactions and the stochastic terms are introduced in order to describe the input that each neuron receives from the rest of the system. An equally sophisticated background of probability theory is the basis of the clustering techniques introduced for genes and protein classification. The necessity to use more complex algorithms comes from the huge amount of data obtained by microarray techniques. In fact this technology is able to extract thousands of genes from the DNA of a cell at the same time and so the problem of collecting the genes in groups (or clusters) with the same function has become crucial. But before arriving at the level of understanding these problems the students, as many students in biology and biotechnology, have to become more acquainted with the usual calculus and probability. Even if they already had an introduction to physics and to calculus, they need other training in order to understand the matter of this book. Thus each chapter is accompanied by appendixes containing the details of some necessary calculus, a brief description of the mathematics or statistics or physics necessary for understanding the meaning of the text contained in the chapters. In other words the chapters contain the description of the model used and the main results without giving all the details of calculations which can be found in the appendixes. There are also solved problems in order to give examples of all the techniques and models introduced. Since mathematical or physical concepts become less abstract if realized with computer programs, an important part of the book is devoted to illustrate the models and their results by means of programs with useful graphic outputs. So the appendixes also contain elementary matlab programs which realize many models and techniques described in the chapters. Part of the exercises are solved by means of Matlab programs. If one wants to compare this book with the usual books describing biological models we can say that this book brings the beginner, the student of the first years of physics, mathematics and biotechnology to the level of understanding the basic of the models used in neurobiology and cluster analysis, giving him also all the necessary instruments and concepts for solving some interesting realistic model or doing cluster analysis. A person or student with this book can build new models or understand research articles on the simulations of neuronal systems. The rest of the literature is more advanced so it is not for the beginners to whom this book is addressed.

Brunello Tirozzi, Daniela Bianchi, Enrico Ferraro

Contents

Appendices   161

PART 1
Neurobiological models

Chapter 1

RC circuit, spiking times and interspike interval

1.1 Introduction

In this chapter we start the discussion of the simplest model of real neu-
rones, the Integrate & Fire model. Because of its simplicity its use is rather
widespread especially for describing the behavior of large systems of interacting neurons. According to this model the neuron is described as a simple
capacitor with an external resistance. The simplicity of this model makes
it useful also from the didactic point of view for explaining the concepts of
spike, spiking time, interspike interval, spiking frequency. Many variants
of the model are introduced showing the connection among the structure
of the neuron and the sequence of spiking times. Typical applications of
this model is the question of synchronous firing of neurons or the synaptic
growth under Hebbian learning.

1.2 Electric properties of a neuron

The neuron is a cell with electric activity. The membrane of the cell has an
electric potential V_m called *membrane potential* and is assumed, as a first
approximation, to be equal at all the points of the membrane, see Fig. 1.1.

The presence of such an electric potential at the membrane of the neuron
is the result of the charges balancing between the internal and external
environment of the cell. Several types of ions of either positive or negative
charge are present outside and inside the cell, and the difference between
the inner and the outer concentration of the diverse ion species produces
the polarization of the membrane. The membrane potential is measured
in Volts (V), the energy E of a charge q at a point P of the membrane at
potential V_m is equal to $E = qV_m$. It represents the work done to bring

3

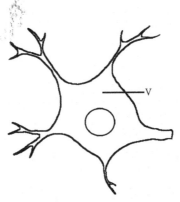

Fig. 1.1   The membrane of the neuron cell is assumed to have constant electric potential V.

the charge q from infinity to the point P where the potential is V_m. If the charge is measured in Coulomb (C), the energy is measured in Joule (J). Therefore, in order to have a charge of 1 Coulomb on the membrane of a neuron whose membrane potential is equal to 1 Volt, it is necessary to do a work of 1 Joule. To understand this order of magnitude, it must be noted that 1 Joule of work corresponds to the work of a force of 1 Newton that, acting on a mass of 1 kg, changes its position of 1 meter, the Newton being defined as the force which gives the acceleration of 1 m/s^2 to the mass of 1 kg. From an electrostatic point of view, the force of 1 Newton corresponds to the repulsion between two equal charges of 1 Coulomb placed at the points P_1 and P_2 respectively, in the empty space at a distance of 1 meter:

$$1 \text{ Newton} = \frac{1}{4\pi\varepsilon_0} \frac{(1 \text{ Coulomb})^2}{(1 \text{ m})^2}$$

where $\varepsilon_0 = 8.85 \times 10^{-12} C^2/\text{Nm}^2$.

The electric activity of a neuron is due to the continuous exchanges of electric currents or of charges with other neurons. The neurons are not isolated but have strong connections among each other through the dendrites, some filaments starting from the membrane of the neuron. See Fig. 1.2.

For example each neuron of the cerebral cortex is connected with $60,000$ other neurons as average number. The charges exchanged during the interactions among each other is of the order of 3×10^{-11} (it is well known that the Coulomb is a very large unit). The flow of the charge among neurons corresponds to an electric current, according to the original definition:

$$I = \frac{Q}{t}$$

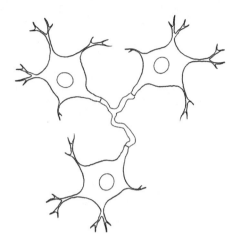

Fig. 1.2   The network formed by neurons is very complicated.

A current of 1 Ampere is obtained in a circuit where 1 Coulomb of charge passes in 1 second. The time a current or a charge takes to go from one neuron to the other has an order of magnitude of 10^{-3} msec so the currents exchanged at neuronal level are of the order of

$$I = \frac{3 \times 10^{-11}}{10^{-3}} \approx 3 \times 10^{-8} \text{ Ampere} = 30 \text{ nA}$$

$$1 \text{ nA} = 10^{-9} \text{Ampere}$$

For example, if the spinal motoneuron of a cat gets 3 nA for 25 msec it gets a charge of the order of

$$Q = 3 \times 10^{-9} \cdot 25 \times 10^{-3} = 7.5 \times 10^{-11} \text{ Coulomb}$$

The potential of the neuron under this current has a behavior of the type shown in Fig. 1.3, we shall derive such a behavior from the models we are discussing in this section.

Its behavior is non-linear in time. The potential increases (depolarization: term used in neurobiology) then decreases (hyperpolarization). The dependence of the potential on the time is non-linear at all time and also the dependence on the input current is non-linear. So we have to look for a non-linear model. What does it mean by non-linear model? It means that the equations associated with the model are non-linear. In the following we will make use of this definition which is of fundamental importance. An equation is linear if any linear combination of two of its solutions is still

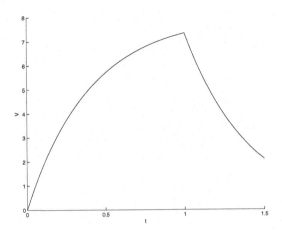

Fig. 1.3   The behavior as a function of time of the potential of a neuron under the input of a constant current acting for a finite time.

a solution of the equation. For example consider an ordinary differential equation of the first order

$$\frac{dy}{dt} + ay = 0$$

This equation is linear because if $y_1(t), y_2(t)$ are two solutions of the equation furthermore $Ay_1 + By_2$ is a solution as one can easily verify.

1.3   Lapicque or I& F model

In the previous section we concluded to the conclusion that we need to construct a model such that $V(t)$ is a non-linear function of the input current $I(t)$. In order to do this let us recall some elementary fact from the theory of electric circuits. The usual connection among electric potential and current is given by the Ohm's law (which can also be used as a definition of the resistance)

$$V = IR$$

The circuit with a resistance $R = 1$ Ohm between two points A and B has a current of 1 Ampere flowing in it if the difference of potential between A and B is 1 volt. A linear behavior means that whatever the input current I is, the electric potential V is proportional to it, see Fig. 1.4.

The slope of the line is just the resistance R. If the relationship between the input current to a neuron and the membrane potential is linear, the

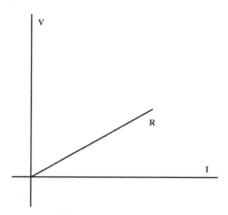

Fig. 1.4   The linear relationship among potential and current.

potential will have a constant value when a constant input current is applied to a neuron for a finite interval of time and should then be equal to zero as one can see from Fig. 1.5.

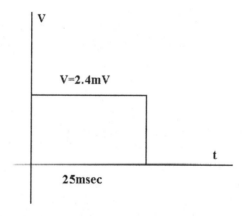

Fig. 1.5   Constant behavior of the electric potential V when a constant current is applied for 25 msec of time.

In this example we take $I = 3$ nA for $t = 25$ ms, $R = 8 \times 10^5$ Ω we have $V = 3 \times 10^{-9} \cdot 8 \times 10^5 = 24 \times 10^{-4}$ Volts $= 2.4$ mV. V is equal to this constant for $t \in (0, 25)$ ms. But in the case of the real neurons the response is non-linear as we can see from Fig. 1.6.

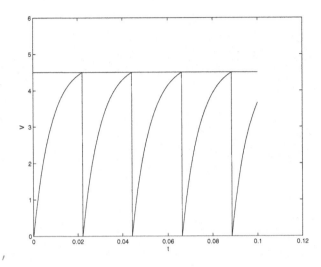

Fig. 1.6   Non-linear behavior of the electric potential V, when the potential reaches the threshold $V0 = 4.5$ V it goes immediately to 0.

In the figure a new quantity $V0$ appears. It is the threshold for the potential. This means that $V(t)$ cannot be larger than $V0$, and, if for some $t^* : V(t^*) = V0$ then the value of $V(t)$ immediately after is zero as one can see from the figure. Note that the input current in this picture is constant as in the previous case and it lasts for 25 msec. When the threshold is reached the neuron emits a spike, i.e. a train of electric current, an electric current of short duration. During this process the neuron emits a charge $Q = CV0$ because the potential of the membrane goes to zero so there are no more charges on it. So if the threshold is $V0 = 60$ mV and $C = 3 \times 10^{-9}$ Farad then $Q \approx 18 \times 10^{-11}$ Coulomb, the charge exchanged among the neurons in an elementary spike is very small. One can make a model of this behavior using the model of Lapicque (1907) or Integrate & Fire model, see [Tuckwell (1988); Feng (2004)]. The model is a RC circuit with external input current, Fig. 1.7.

The membrane potential of an isolated neuron tends to hyperpolarize, so the neuron can be represented as an RC circuit. The capacity C of a capacitor is the constant relating the potential V with the charge Q of a conductor

$$Q = CV$$

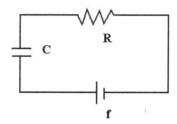

Fig. 1.7   Elementary RC circuit as a neuron model.

If $Q = 1$ Coulomb and $V = 1$ Volt, then $C = 1$ Farad. The RC circuit is described by the following linear differential equation with constant coefficients (i.e. the simplest one):

$$\frac{dQ}{dt} + \frac{V}{R} = 0 \Longrightarrow C\frac{dV}{dt} + \frac{V}{R} = 0$$

The term $C\frac{dV}{dt}$ measures the decrease of the charge Q of the capacitor in the unit time

$$\frac{dQ}{dt} = \frac{d(CV)}{dt} = C\frac{dV}{dt}$$

so it is equal to $-I$, the current flowing in the circuit. The current flowing through the resistance R is, by the Ohm's law,

$$I = \frac{V}{R}$$

Putting all the information together we get the equation of the *RC* circuit:

$$\frac{dQ}{dt} = -I = -\frac{V}{R}$$

$$C\frac{dV}{dt} + \frac{V}{R} = 0$$

$$\frac{dV}{dt} + \frac{V}{RC} = 0$$

The constant *RC* has the dimension of a time:

$$[RC] = \frac{[V]}{[I]}\frac{[Q]}{[V]} = \frac{[Q]}{[I]} = [t]$$

Therefore the equation can be rewritten as

$$\frac{dV}{dt} + \frac{V}{\tau} = 0$$

where $\tau = RC$ is the characteristic time of the potential. The simple method for solving this equation is shown in Appendix A. From this method and the general properties of differential equations it follows that the solution is uniquely determined by the initial value of the potential $\overline{V}$. Thus the problem to solve is the RC circuit.

Definition 1.1. *RC circuit*
    The equation for the potential of the capacitor of an RC circuit is

$$\begin{cases} \frac{dV}{dt} + \frac{V}{\tau} = 0 \\ V(0) = \overline{V} \end{cases}$$

and the solution is given by

$$V(t) = \overline{V}e^{-\frac{t}{\tau}} \tag{1.1}$$

From this solution it is clear why τ is considered a characteristic time of the neuron, it simply equals to the time it takes, for a neuron without external input, to lower the initial potential $\overline{V}$ by a factor e^{-1}. From Fig. 1.7 one can also find an interesting interpretation of the voltage f of the battery. In this case we have to include another term in the equation

Lemma 1.1. *RC circuit with battery*
    The equation of an RC circuit with a battery f is

$$\frac{dV}{dt} + \frac{V}{\tau} = -\frac{f}{\tau}, \quad V(0) = \overline{V} \tag{1.2}$$

Proof.
    The equation for V is found by making elementary arguments:

$$V + f = IR \tag{1.3}$$

$$I = -\frac{dQ}{dt} \tag{1.4}$$

$$-\frac{dQ}{dt} = -C\frac{dV}{dt} \tag{1.5}$$

$$V + f = -\tau\frac{dV}{dt} \tag{1.6}$$

$\square$

The solution of the equation (1.2) is $V(t) = \overline{V}e^{-\frac{t}{\tau}} - f(1 - e^{-\frac{t}{\tau}})$ so for times $t \gg \tau$ the potential is $V \sim -f$, in other words, $-f$ represents the resting potential of the membrane. The derivation of the solution can be found in appendix A. Since the neuron is not isolated but connected with a very complicated network of neurons, the model must be completed by inserting an external current $I(t)$ which can be a difficult function of time and also have a random nature. This is a consequence of the fact that the spiking activity of a large system of neurons is a random event so the charges from the other neurons arrive at random. Thus the full model of the neuron of Lapicque is

Definition 1.2. *Integrate & Fire model (I&F)*

$$\begin{cases} \dfrac{dV}{dt} + \dfrac{V}{\tau} = \dfrac{I}{C} \\ V(0) = \overline{V} \\ \text{if} \quad \exists t : V(t) = \theta \to V(t^+) = 0 \end{cases}$$

where $t^+ = t + \epsilon$ with ϵ very small, so is the instant of time just after the spiking time.

Definition 1.3.

The times $t_1, ..., t_n$ at which the condition $V(t) = \theta$ is reached, are a random sequence and are called i.e. the spiking times and so also the differences $T_1 = t_2 - t_1$, $T_2 = t_3 - t_2$,... are random. The difference T_i is called *interspike interval*.

The interspike interval is one of the main objects of experimental and theoretical investigation in neurobiology since the probability distribution of T_i characterizes the state of the neuron in a certain biological situation. Since the T_i are random it is the case to start to introduce the empirical average of interspike intervals.

Definition 1.4.

Given N measures of T, $T_1, ..., T_N$, the empirical mean of T, named ET, is

$$ET = \frac{1}{N}(T_1 + ... + T_N) \tag{1.7}$$

The inverse of the average interspike interval gives a measure of the spiking frequency:

$$f = \frac{1}{ET} \tag{1.8}$$

The general definition of the spiking frequency is

Definition 1.5.

$$f = \frac{\text{number of spikes}}{\text{unit of time}} \tag{1.9}$$

i.e. f is defined as the number of spikes a neuron emits in the unit of time. This is also a fundamental definition since the spiking frequency characterizes the state of the neurons of a given system. Usually a system of neurons has two states: a normal state of low activity and a state of high activity corresponding to the functioning that the system has to accomplish. For example the neurons responsible of the emission of the oxytocin, a neurotransmitter responsible of parturition and suckling, when they are in a normal state have a spiking frequency $f \sim 5 - 10 \frac{\text{spikes}}{\text{second}}$ and when they are acting they jump to a *bursting* state of approximately $f \sim 70 - 80 \frac{\text{spikes}}{\text{second}}$. A big problem is to understand if such states of activity of the neurons are synchronized or not. Synchronization means that the neurons reach the threshold $V0$ at the same time, the question is important since the global signal exiting from a neural system is stronger in such a case. Probably also the Parkinson disease is connected with a synchronized activity of motor neurons and epilepsy is a synchronized state of the neuron of the cerebral cortex. All these important questions can be investigated also by simulating the behavior of a neural system with the Integrate & Fire model, the easiest model to be used.

Chapter 2

Calculation of interspike intervals for deterministic inputs

2.1 Introduction

In the first chapter we have introduced the main variables and parameters which characterize the neural activity and are currently measured in neurobiological experiments. We have also introduced the Integrate & Fire model as the simplest model adapt to describe the behavior of a system of neurons. The idea is to treat a neuron like a simple RC circuit with the condition that if the electric potential $V(t)$ reaches a certain threshold then it emits a spike and suddenly the potential drops to zero. We are going to analyze in detail such a model in this chapter and to compute the solutions of the simple differential equations associated with this system. The details of the derivations of the solutions are given in appendix B. We also show graphs representing the solution which have been produced solving numerically the model using some Matlab programs which are also explained in appendix B.

2.2 Case of constant input current

Before solving the model in the case of a constant input current arriving at the neuron, let us discuss again the model and the neuron in more detail. The structure of the membrane of the neural cell is not so simple as described in the previous section. There are many filaments arriving at a neuron from the other neurons called dendrites, and when the neuron emits a spike, the current or charge flows away from it along a larger tubular membrane called axon, the axon divides itself into many smaller filaments and so the charge flows also along them arriving at the other neurons, Fig. 2.1.

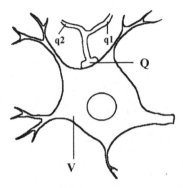

Fig. 2.1   Schematic representation of the neural connections, $Q = q_1 + q_2$.

The charge exiting from the axon divides itself among the different dendrites according to the usual Kirchhoff law of the currents. So the charge flowing in the axon divides itself in the dendrites and smaller charges arrive at the neurons connected with the spiking neuron. The arrival of the charge on the membrane of a neuron generates a sudden change of its potential. If C is the capacity of the neuron and q is the charge arriving at its membrane, its potential will change the quantity:

$$V \to V + \frac{q}{C}.$$

If θ is the threshold of the neuron, the charge emitted by it during the spike is

$$Q = \theta C$$

since the potential at the moment of the emission of the spike is θ and after the spike is zero so all the electric potential contributes to the charge. Let us now solve the model for an external input current. So we have to solve the following system:

$$\frac{dV}{dt} + \frac{V}{\tau} = \frac{I}{C} \quad V(0) = 0 \quad \text{if} \quad V(t) = \theta \to V(t^+) = 0 \qquad (2.1)$$

and find the conditions of spiking and the value of the interspike time. We will show that there is a spike if $IR \geq \theta$. We show in Fig. 2.2 the graph of the potential as a function of time for the choice of the constants $I = 60 \times 10^{-9}$, $R = 10^9$, $C = 10^{-11}$, $\theta = 5$ which have been chosen in such a way that the threshold condition is satisfied.

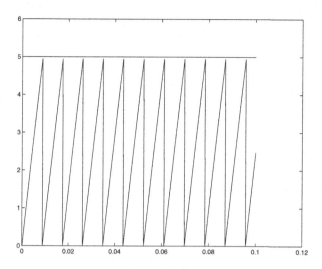

Fig. 2.2 Sequence of spikes of the neuron under a constant input current.

Lemma 2.1. *Consider a neuron described by an I&F model with constant input I, threshold θ, $\tau = RC$. Its potential $V(t)$ satisfies the equations*

$$\begin{cases} \dfrac{dV}{dt} + \dfrac{V}{\tau} = \dfrac{I}{C} \\ V(0) = 0 \\ if \quad V(t) = \theta \rightarrow V(t^+) = 0 \end{cases}$$

The solution is

$$V(t) = IR(1 - e^{-\frac{t}{\tau}}) \qquad (2.2)$$

If $\theta < IR$ the neuron emits periodic spikes with interspike interval

$$T = \tau \log\{\frac{1}{1 - \frac{\theta}{IR}}\} \qquad (2.3)$$

and the solution repeats periodically as in Fig. 2.2.

Proof.

The solution of this equation can be found exactly with the method used in appendix A:

$$V(t)e^{\frac{t}{\tau}} = \frac{I}{C}(e^{\frac{t}{\tau}} - 1)$$

where the variable $\overline{V}$ has been put equal to zero because the initial condition is zero and $-\frac{t}{\tau}$ has been substituted by $\frac{I}{C}$, we arrive at the final expression

$$V(t) = IR(1 - e^{-\frac{t}{\tau}}) \tag{2.4}$$

So there will be a spike if $IR \geq \theta$ and the interspike interval T can be computed solving the equation

$$IR(1 - e^{-\frac{T}{\tau}}) = \theta$$

The solution is easily found inverting the exponential by means of the logarithm:

$$T = \tau \log \left[\frac{1}{1 - \frac{\theta}{IR}} \right] \tag{2.5}$$

which makes sense only because $\theta < IR$, otherwise the argument of the logarithm would be negative. $\qquad\qquad\qquad\qquad\qquad\qquad\qquad\qquad$ □

In Fig. 2.2 the graph of the solution as a function of t is displayed, in the graph it is clear that when the potential reaches the threshold $\theta = 5$ it goes back immediately to zero as requested from the model. This graph has been obtained integrating numerically the equation of the model with a method shown in appendix B and the method has been implemented using a Matlab program again shown in the same appendix. The time interval between two subsequent points where the potential is equal to the threshold is the interspike interval, it can be computed by using the numerical algorithm as well as by means of formula (2.5), actually they coincide, it gives a useful way to check the numerical algorithm. The necessity of introducing the numerical algorithm comes from the fact that the models of real neurons are not so simple as this one and so an analytic solution cannot be found. The program which solves these more complicated cases can be found by making an easy generalization of the programs show in the appendix B. Let us make explicitly this check. Let us substitute the numerical values in (2.5): $I = 60 \times 10^{-9}$, $R = 10^9$, $C = 10^{-11}$, $\theta = 5$. Thus $\tau = 10^{-2}$, using the formula of the interspike time (2.5) we get the following estimate for $T \sim 0.0087011$ while the program in appendix B gives just $T \sim 0.0087$. The difference is very small and so the numerical approximation made with the method of appendix B is rather good. In appendix B we give the details of the algorithm and the corresponding program.

2.3 Constant input current for a finite time

Now let us consider the case in which a constant input current I is applied for a finite time interval u, as in Fig. 2.3.

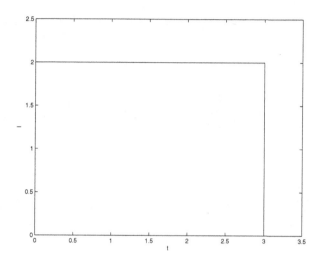

Fig. 2.3 A constant input current $I = 2$ applied for $u = 3$ seconds.

The system corresponding to this case is of the type

$$\frac{dV}{dt} + \frac{V}{\tau} = \frac{I}{C} \quad \text{for} \quad t < u \quad V(0) = 0 \tag{2.6}$$

$$\frac{dV}{dt} + \frac{V}{\tau} = 0 \quad \text{for} \quad t > u \quad V(u) = IR(1 - e^{-\frac{u}{\tau}}) \tag{2.7}$$

The system has a form which depends on the considered time interval, if $t < u$ there is an input current I otherwise the current is zero and so the r.h.s. is zero for $t > u$. The condition at time $t = u$ on the potential $V(t)$ is the value taken by the solution in the time interval $t < u$ $V(t) = IR(1 - e^{-\frac{t}{\tau}})$ at the instant $t = u$. We now establish the condition for firing and find the solution and the interspike time.

Lemma 2.2.

Consider a neuron described by an I&F model with a constant input I applied for a finite time u (Fig. 2.3), threshold θ, $\tau = RC$. The potential V is the solution of the equations

$$\begin{cases} \dfrac{dV}{dt} + \dfrac{V}{\tau} = \dfrac{I}{C} \\ \quad for \quad t < u \\ \qquad V(0) = 0 \\ \dfrac{dV}{dt} + \dfrac{V}{\tau} = 0 \quad for \quad t > u \\ V(u) = IR(1 - e^{-\frac{u}{\tau}}) \end{cases}$$

Let t_1 be the spike time, $0 < t_1 < u$, then solution of the system is:

$$\begin{cases} V(t) = IR(1 - e^{-\frac{t}{\tau}}) \quad for \quad 0 < t < t_1 \\ V(t) = IR(1 - e^{-\frac{(t-t_1)}{\tau}}) \quad for \quad t_1 < t < u \\ V(t) = IR(1 - e^{-\frac{(u-t_1)}{\tau}})e^{-\frac{(t-u)}{\tau}} \quad for \quad u < t < u + v \end{cases}$$

The interspike time is:

$$T = \tau \log \left[\dfrac{1}{1 - \dfrac{\theta}{IR}} \right] \tag{2.8}$$

the condition for spiking are $\theta \le IR$ and $\tau \le u$. If τ/u is small there might be many spikes, if $\tau > u$ there are no spikes.

Proof.

The solution of the first equation, for $0 \le t \le u$ is like the solution found before when I is constant for any time and so is of the type $V(t) = IR(1 - e^{-\frac{t}{\tau}})$. In the second expression there is the exponential decay of the potential due to the absence of input current but with a decay factor $e^{-\frac{(t-u)}{\tau}}$ which depends on the difference $t - u$ since the origin of the time is in u for the second equation. The condition for having a spike is as above $V(t) > \theta$ and clearly it can be satisfied only for $t < u$ since after the time u the potential is always decreasing. Thus we get the same expression as before, but under the condition that $T < u$. If τ is smaller than u then this condition is satisfied, since τ is the characteristic time for which the potential $V(t) = IR(1 - e^{-\frac{t}{\tau}})$ reaches its maximum value and if $\tau << u$ there might be many spikes since the maximum value of v can be reached many times. If $\tau > u$ then the potential approaches the limit value too slowly and it can also start to decrease for $t > u$ and so there no spikes at all. $\qquad\square$

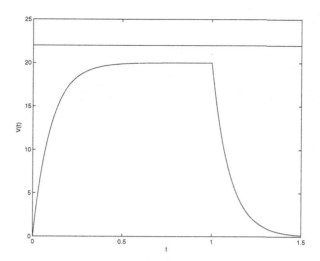

Fig. 2.4   Behavior of the potential when the threshold $\theta = 22$ mV is not reached.

In Fig. 2.4 it is represented by the case when the threshold is not reached.

In Fig. 2.5 it is represented by the case in which the condition $\tau < u$ is satisfied, $u = 1$, $\tau = 0.1$, so there are many spikes and for $t > 1$ there is the exponential decay as expected.

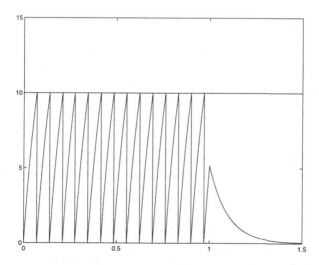

Fig. 2.5   Multiple spikes in the case when the characteristic time τ is less than u.

2.4   Constant input current with a periodic pattern

We now analyze the case of an impulsive current, i.e. a current taking a constant value I for a certain interval of time u and equal to zero for another interval of time v. In Fig. 2.6 an example of this input is shown for a finite time. There are many spiking patterns, in this lemma we analyze some of them.

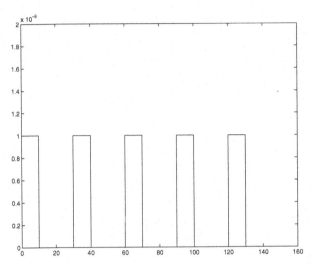

Fig. 2.6   A current of $I = 10^{-9}$ Ampere applied for $u = 10$ sec and equal to zero for $v = 20$ sec. The pattern is periodic and in the figure only 5 impulses are shown.

Lemma 2.3.

In the case of a neuron described by an I&F model with periodic inputs of the type of Fig. 2.6 with parameters u and v, threshold θ, $\tau - RC$ there are many different spiking patterns:

(1) *If $IR(1 - e^{-\frac{u}{\tau}}) > \theta$ the neuron can emit spikes.*

(2) *If $RC << v$, $RC \sim u/2$, there is a periodic emission of spikes, the interspike time is $T = u + v$, the first spike happens at t, $0 \leq t \leq u$. (Fig. 2.7).*

(3) *If $RC >> u+v$ there is a periodic emission of spikes with longer interspike interval as in Fig. 2.8, in that case for example $T = 3u + 2v$.*

(4) *If $RC << u, v$, there are many spikes in the intervals $(n(u + v), n(u + v) + u)$ then the neuron is passive for v seconds and the patterns repeat*

periodically, there are two interspike times T_1 for the high frequency of emission of spikes and $T_2 = T_1 + v$ for the slow emission (Fig. 2.9).

Proof.

We already know that the potential for a constant input which lasts for a finite time interval u is

$$\begin{cases} V(t) = IR(1 - e^{-\frac{t}{\tau}}) & \text{for} \quad 0 < t < u \\ V(t) = IR(1 - e^{-\frac{u}{\tau}})e^{-\frac{(t-u)}{\tau}} & \text{for} \quad t > u \end{cases}$$

The condition for the emission of a spike at time $t_1 < u$ is $IR(1 - e^{-\frac{t_1}{\tau}}) \geq \theta$, which holds because $IR(1 - e^{-\frac{u}{\tau}}) > \theta$ since $u > t_1$. If this condition holds, the potential at time $t_1 + \epsilon$ is zero and so there will be another spike if $u - t_1 \geq t_1$ and one could apply this argument repeatedly. In the case when there is only a spike in the interval $(0, u)$ the potential has the following behavior:

$$\begin{cases} V(t) = IR(1 - e^{-\frac{t}{\tau}}) & \text{for} \quad 0 < t < t_1 \\ V(t) = IR(1 - e^{-\frac{(t-t_1)}{\tau}}) & \text{for} \quad t_1 < t < u \\ V(t) = IR(1 - e^{-\frac{(u-t_1)}{\tau}})e^{-\frac{(t-u)}{\tau}} & \text{for} \quad u < t < u + v \end{cases}$$

The second expression is due to the fact that the potential starts from zero after the spike at time $t = t_1$. If $\tau << v$ the potential is zero at the time $t = u + v$ and so the pattern of spikes is repeated. Thus we have a repetitive firing pattern something which happen very often in real neural system. The different behavior depends on the ratio $\tau/(u + v)$. In Fig. 2.7 the case with one spike in the interval $(n(u+v), n(u+v)+u)$ is shown, in this case the ratio τ/u is of the order one, the interspike interval is $T = u + v$.

In Fig. 2.8 the activity pattern of a neuron with τ bigger than $u + v$ is shown, in this case τ is such that the spike happens only at the end of the third interval of length u, i.e. for $t_n = 2n(u + v) + u$. It is interesting to notice also that there are subthreshold oscillations which is also a widely diffuse phenomenon in the neuronal dynamics. In this case the interspike interval is $T = 3u + 2v$.

Finally in the last figure of this section, Fig. 2.9, the case when $\tau = u/10$ is shown, there are many spikes in the interval $(n(u + v), n(u + v) + u)$ and when the interval $(n(u + v), n(u + v) + u)$ is finished the potential goes to zero. In this case there are two interspike intervals: $T_1 = t_n - t_{(n-1)}$ where t_n are the spiking times when the current is present and $T_2 = T_1 + v$. $\quad\square$

Notice that the transition of a neuron going from a low activity rate Fig. 2.7, Fig. 2.8 to a high activity rate Fig. 2.9 is a usual event in the

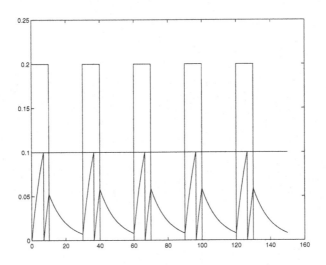

Fig. 2.7   Case $u - t_1 \leq t_1$ such that only one spike takes place in the interval $(0, u)$.

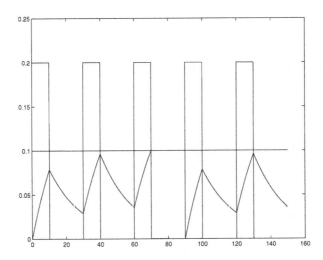

Fig. 2.8   Case when $\tau \gg u + v$ and there are few spikes and subthreshold oscillations.

neural dynamics which corresponds to a change of the function of the neural system. Such transition can be seen for example in the system of oxytocin cells when the neurons start to emit the oxytocin, a neurotransmitter which is also a hormone necessary for the lactation and parturition.

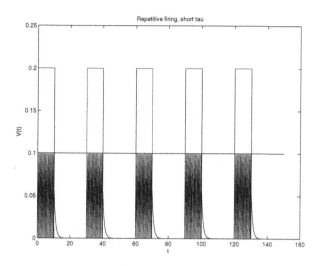

Fig. 2.9   Case of multiply repetitive spikes.

2.5   Periodic instantaneous inputs

Now let us consider a sequence of spikes arriving at a neuron at regular times $T, 2T, 3T, \ldots$ as in Fig. 2.10 where a sequence of 10 such impulses is represented. The duration of the spike is infinitesimal, one could consider the current as a sum of delta functions $q\delta(t - nT)$ but this does not change our analysis. q is the charge that arrives at the neuron in each spike and so it causes an increase of the potential of the neuron of the quantity $k = q/C$ where C is the capacity of the neuron.

We look for the condition of having a spike and what is the interspike interval, where the threshold is θ as usual.

Lemma 2.4. *([Tuckwell (1988)])*

Consider a neuron described by an I&F model with $RC = \tau$, threshold θ and periodic instantaneous inputs arriving at times $t_n = nT$, (Fig. 2.10), $k = q/C$, $V(0) = 0$. Then

(1) If $k > \theta$ the neuron emits spikes at times t_n.

(2) If $k < \theta$ the potential $V(t)$ of the neuron oscillates, asymptotically, among two values $V_{min} \leq V(t) \leq V_{max}$ with $V_{min} = ka\frac{1}{1-a}$, $a = e^{-\frac{T}{\tau}}$, and $V_{max} = k\frac{1}{1-a}$. The spiking condition is $V_{max} > \theta$ and interspike interval is NT, where N is the number of steps needed to reach the asymptotic regime.

Proof.

(1) In the case $k > \theta$ the argument is straightforward because the potential of the neuron starts from the zero value and increases abruptly of a quantity k whenever it receives the impulse. In this case the threshold is crossed and the neuron emits a spike and then its potential goes back to zero again and this situation repeats indefinitely. So the spiking times are $t_n = nT$.

(2) In the case $k < \theta$ we have to make a more complicate argument. Let us start our analysis at the time $t = 0$. We have the simple behavior in the interval $(0, T)$:

$$\begin{cases} V(t) = 0 & \text{for} \quad 0 < t < T \\ V(t) = k & \text{for} \quad t = T \end{cases}$$

In the interval $(T, 2T)$ the electric potential $V(t)$ will decay exponentially since there is no external current or input so it satisfies the following equation

$$\begin{cases} \frac{dV}{dt} + \frac{V}{\tau} = 0 \\ V(T) = k \end{cases}$$

thus we have $V(t) = ke^{-\frac{(t-T)}{\tau}}$ for $t \in (T, 2T)$. At an instant t just before $2T$, $t = 2T^- \equiv 2T - \epsilon$, the potential has the value $V(2T^-) = ke^{-\frac{T}{\tau}}$. At the time just after the arrival of the spike at $t = 2T$ the value of the potential

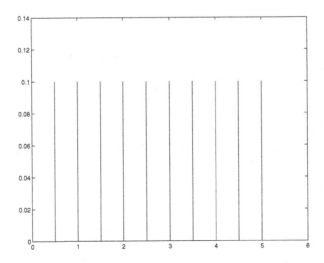

Fig. 2.10   Periodic impulses.

is $V(2T^+) = ke^{-\frac{T}{\tau}} + k = k(1 + e^{-\frac{T}{\tau}})$. In the next interval $(2T, 3T)$ the potential satisfies the following system:

$$\begin{cases} \frac{dV}{dt} + \frac{V}{\tau} = 0 \\ V(T) = k(1 + e^{-\frac{T}{\tau}}) \end{cases}$$

So $V(t) = k(1 + e^{-\frac{T}{\tau}})e^{-\frac{(t-2T)}{\tau}}$ in the interval $(2T, 3T)$. Repeating the same argument n times we arrive at the conclusion that

$$\begin{cases} V(nT^-) = ke^{-\frac{T}{\tau}}(1 + e^{-\frac{T}{\tau}} + e^{-\frac{2T}{\tau}} + \cdots + e^{-\frac{(n-1)T}{\tau}}) \\ V(nT^+) = V(nT^-) + k = k(1 + e^{-\frac{T}{\tau}} + e^{-\frac{2T}{\tau}} + \cdots + e^{-\frac{nT}{\tau}}) \end{cases}$$

We recognize easily that these sums are the power series $1 + a + a^2 + \cdots + a^n$ with $a = e^{-\frac{T}{\tau}}$, so using the well known formula

$$\frac{1 - a^n}{1 - a} = 1 + a + \cdots + a^{n-1}$$

we get, taking into account that $a < 1$,

$$\begin{cases} V(nT^-) = ka\frac{1-a^n}{1-a} \to_{n \to \infty} ka\frac{1}{1-a} \equiv V_{\min} \\ V(nT^+) = V(nT^-) + k = ka\frac{1-a^n}{1-a} + k \to_{n \to \infty} ka\frac{1}{1-a} + k = k\frac{1}{1-a} \equiv V_{\max} \end{cases}$$

In words we can express this result by saying that asymptotically the potential of the neuron will oscillate among a minimal value and a maximal value $V_{\min} < V(t) < V_{\max}$. At each arrival of the spike at time $t = nT$ it will jump from $V_{\min}$ to $V_{\max} = V_{\min} + k$. If the spiking condition $V(t) > \theta$ does not hold then the potential will decrease exponentially in the interval $(nT, (n+1)T)$ with the exponential decay $V(t) - V_{\max}e^{-\frac{(t-nT)}{\tau}}$, since there is no external current in the intervals $(nT, (n+1)T)$, till it gets a new increment k at time $t = (n+1)T$. It is evident that the spiking condition is $V_{\max} > \theta$. If this condition holds we will have, as soon as the potential arrives at this asymptotic regime, that the spike of the neuron will take place at the instant $t = NT$, for N large enough in such a way that the asymptotic regime is reached. After this spike has occurred the neuron starts again from the zero potential and the evolution of the potential will be the same, so that the next spike will take place only after other N impulses arrived at it.

$\square$

2.6   Exercises

In this section we propose some useful exercises for training, the solutions can be found in appendix B. There are also exercises without answers that the reader has to do by himself after having done the exercises for training.

Exercise 1

Consider a neuron described by an Integrate & Fire model with a scheme as in Fig. 1.7 and values $C = 10^{-11}$ Farad, $R = 10^6$, $f = 60$ mV.

(a) Find the rest potential of the neuron. A: $V_{rest.} = -60$ mV.
(b) Find the time necessary for the neuron to reach the rest potential starting from the initial potential $V(0) = 10$ mV. A: $t = \tau = 10^{-2}$ msec.
(c) Suppose that a charge $q = 10^{-12}$ Coulomb arrives from outside to the neuron, what is the change of potential? A: $\Delta V = 10^{-1}$ volt.
(d) How much the resistance R has to change in order that the time of the question (b) decreases of a factor 10? A: $R = 10^5$.

Exercise 2

Suppose that the above neuron receives a constant current $I = 10$nA $= 10^{-8}$ Ampere and that $C = 3 \times 10^{-9}$ Farad, $R = 10^6$ Ohm, $f = -70$ mV.

(a) Find the time evolution of the potential. A: $V(t) = (IR - f)(1 - e^{-\frac{t}{\tau}})$.
(b) Find if the condition of having a spike is satisfied if threshold is $\theta = 60$ mV. A: $(IR + f) > \theta$.
(c) Suppose that $f = 0$, how much must the current I increase in order that the spiking condition still holds? A: $\Delta I = 50$ nA.
(d) Find the interspike interval using the current $I = 70$ nA. A: $T = 1.5$ msec.
(e) Find the current to send to the neuron in order that the interspike interval becomes twice bigger. A: $I = 0.96$ nA.

Exercise 3

Consider $f = 0$, the input current defined by

$$\begin{cases} I = I & 0 \leq t \leq u \\ I = 0 & t > u \end{cases}$$

$\tau = 2.4$ msec, $I = 10^{-9}$ A, $R = 10^6$ Ohm, $\theta = 60$ mV.

(a) Find u in such a way that the spike condition holds for $t = u$. A: $u = 2.2$ msec.

(b) Find the amount of charge emitted by the neuron during a spike. A: $q = 144 \times 10^{-12}$ Coulomb.

(c) Suppose that a neuron A emits such the same charge q every $T = 4.8$ msec to another neuron B of capacity $C = 10^{-9}$ Farad. What are the spiking times t_n of the neuron B? A: $t_n = nT$.

(d) Consider a sequence of periodic inputs as in Fig. 2.6. Find the value of I, R, u and v such that the situation of the question (c) takes place. A: $I = 10^{-7}$ Ampere, $R = 10^6$ Ohm, $\theta = 60$ mV, $u = 2.2$ msec, any v.

(e) Find the spiking times. A: $t_n = t_1 + n(u + v)$, where $IR(1 - e^{-\frac{t_1}{\tau}}) = \theta$.

Exercise 4

Consider a neuron which receives a current of the type:

$$\begin{cases} I = I_1 & 0 \le t \le u \\ I = I_2 & u \le t \le u + v \end{cases}$$

with initial potential $V(0) = 0$ and $I_1 < I_2$.

(a) Find the condition that the neuron emits at least one spike in the interval $(0, u)$

(b) Suppose that the neuron has emitted no spike in the interval $(0, u)$ find the condition that it emits a spike in the interval $(u, u + v)$

A:

(a) $I_1 R(1 - e^{-\frac{u}{\tau}}) > \theta$

(b) $R[I_1(1 - e^{-\frac{u}{\tau}}) - I_2]e^{-\frac{v}{\tau}} + I_2 R > \theta$.

Exercise 5

Find the condition for an Integrate & Fire model to have the same output as in the previous exercise when it receives a periodic input current of the type:

$$\begin{cases} I = I_1 & n(u + v) \le t \le n(u + v) + u, & n = 0, 1... \\ I = I_2 & n(u + v) + u \le t \le (n + 1)(u + v), & n = 0, 1... \\ I = 0 & (n + 1)(u + v) \le t \le (n + 1)(u + v) + w & n = 0, 1... \end{cases}$$

supposing that $u, v, w >> \tau$, $\tau = RC$, $I_1 < I_2$, $V(0) = 0$. This is the current of the previous example repeated at periodic intervals and the impulses are separated by an interval of time w. A:

$$I_1 R > \theta \quad n(u+v) \leq t \leq n(u+v) + u \quad n = 0, 1... \qquad (2.9)$$

$$I_2 R > \theta \quad n(u+v) + u \leq t \leq (n+1)(u+v). \quad n = 0, 1... \qquad (2.10)$$

Exercise 6

In the previous exercise find u such that there is only one spike at the time t_1 in the intervals $n(u+v) \leq t \leq n(u+v) + u$ with $t_1 \sim n(u+v) + u$ and only one spike the intervals $n(u+v) + u \leq t \leq (n+1)(u+v)$. A:

$$\begin{cases} \tau \log \frac{1}{1 - \frac{\theta}{I_1 R}} \sim u & n(u+v) \leq t \leq n(u+v) + u \quad n = 0, 1... \\ \tau \log \frac{1}{1 - \frac{\theta}{I_2 R}} > \frac{v}{2} & n(u+v) + u \leq t \leq (n+1)(u+v). \quad n = 0, 1... \end{cases}$$

Exercise 7

Find the spiking condition for a neuron with relaxation time $\tau = RC$ such that there is only spike at time t_1 in the interval $0 < t < u$ and one spike in the interval $u < t < u + v$ with an input current of the type:

$$\begin{cases} I = I\frac{t}{\tau} & 0 \leq t \leq u \\ I = I\frac{u}{\tau} & u \leq t \leq u+v \end{cases}$$

A: $IR(\frac{u}{\tau} - (1 - e^{-\frac{u}{\tau}})) > \theta$ for $0 \leq t \leq u$ and $IR(\frac{u-t_1}{\tau} - (1 - e^{-\frac{u-t_1}{\tau}}))e^{-\frac{v}{\tau}} + IR\frac{u}{\tau}(1 - e^{-\frac{v}{\tau}}) > \theta$ for $u \leq t \leq u + v$.

Exercise without answers.

Exercise 8
Find the spiking conditions for an Integrate & Fire model with given τ, R, I, θ and with the following input currents:

(a)

$$\begin{cases} I = I\frac{t}{\tau} & 0 \leq t \leq u \\ I = I\frac{u}{\tau}(1 - \frac{t-u}{v}) & u \leq t \leq u+v \end{cases}$$

(b)

$$\begin{cases} I = I_1 & 0 \leq t \leq u \\ I = I_2 & u \leq t \leq u+v \\ I = I_1 & u+v \leq t \leq u+v+w \end{cases}$$

(c)

$$\begin{cases} I = I\frac{t}{\tau} & 0 \le t \le u \\ I = I\frac{u}{\tau} & u \le t \le u+v \\ I = I\frac{u}{\tau}(1 - \frac{t-(u+v)}{w}) & u+v \le t \le u+v+w \end{cases}$$

(d)

$$\begin{cases} I = I\frac{t}{\tau} & 0 \le t \le u \\ I = I\frac{u}{\tau} & u \le t \le u+v \\ I = I\frac{u}{\tau}(1 + \frac{t-(u+v)}{w}) & u+v \le t \le u+v+w \end{cases}$$

Chapter 3

The Fitzhugh-Nagumo and Hodgkin-Huxley models

3.1 Introduction

The $I\&F$ model explained and discussed in detail in the previous chapters is very instructive and gives many interesting characteristics of the neural dynamics. For example one can also get repetitive spiking activity for a neuron with constant input. But the limit is that there is not so much phenomenology of the neuron in the $I\&F$ model since the connection of the depolarization and hyperpolarization of the membrane potential with the incoming or outgoing ionic currents is completely ignored. Instead these currents play a key role in the dynamic of the spike. The neural membrane in fact divides the internal environment of the neural cell from the exterior and there is a well defined distribution of ions in these environments. There is in fact the K^+(potassium), Na^+(sodium), Ca^{--} (calcium), diluted everywhere inside and outside the neural cell but with different concentrations outside and inside the cell. The difference of ionic concentrations outside and inside the cell determines a value of the membrane potential from well known electrostatic equilibrium equations. The arrival of the spike from other neurons causes the opening of the ionic channels in such a way that the ions can go through their respective channels. The ionic channels on the membrane are selective, and the change of the difference of ionic concentrations causes a change of the potential and the formation of a spike. Many features of the potential as a function of time are connected with the ionic dynamic and the $I\&F$ model cannot describe these more complicated and rich phenomena. In this chapter we discuss two models which include such phenomenology at different level of detail and complexity. The dynamic of the spike is reconstructed from these two models and we use some non-linear mathematical analysis to give an interesting explanation of the spiking activity.

3.2   The Fitzhugh-Nagumo model and the general properties of differential equations

As we said before the action potential is created by the dynamics of ionic currents. The concentration of ions plays a key role in the depolarization and hyperpolarization of the membrane potential. The Na^+ concentration inside the neuron is 49 mmole and 460 mmole outside the neurons of the squid. For the concentration of the K^+ the contrary holds: it is 150 mmole inside the neuron and 5.5 mmole outside. Thus the sodium current is going inside the neuron while the potassium current flows outside the neuron. The ions flow in channels which open or close under the action of special proteins called neurotransmitters. The neurotransmitters are sent to the channels from the action of the spikes arriving at the pre-synaptic junction. The channels are selective, they allow only one type of ion to flow through it, so there is the sodium channel and potassium channel.   The relationship among the ionic electric current going through a channel and the electric potential cannot follow the Ohm's law since there are the non linear phenomena of closure and opening of the channel. Thus we have in general to model with a non-linear function $g(V)$ the conductance of the channel and write the following equation for the ionic current:

$$j(V) = g(V)(V - V_a) \tag{3.1}$$

we note that the conductance is the inverse of the resistance, V_a is the inversion potential, i.e. the value of the potential such that the current changes direction. A positive current can be considered, by convention as a current flowing into the neuron and a negative current just the opposite case. The simplest model which includes the dynamic of the current is the Fitzhugh-Nagumo model [Fitzhugh (1960)], [Scott (1975)]. In such a model the sodium current has a conductance of the type

$$g(V) = V(1 - V) \tag{3.2}$$

and the inversion potential V_a is written as a and is a constant without a precise value. The potassium current is not really present in this model but another current $j = -W$ is inserted which allows the model to have the periodic solutions which are characteristic of the spike activity. It can be considered as a preliminary form of the potassium current. W is a variable independent of V and the current $j = -W$ is called the *recovery current*. The definition of the model is

Definition 3.1.

The Fitzhugh-Nagumo model (FN model) is based on two variables $(V(t), W(t))$ which are solutions of the following system of differential equations

$$\begin{cases} \frac{dV}{dt} = V(1-V)(V-a) - W + I \\ \frac{dW}{dt} = b(V - \gamma W) \end{cases}$$

where I is the external current and $b = 1/10$ is a small coefficient.

Remark 3.1. It is important to stress the particular features of this model since they have biological relevance. First of all, different from the $I\&F$ model, the spiking condition is not imposed explicitly so the model will give, under conditions we show afterwards, the sequence of spikes by its internal mechanism not by externally imposed conditions. The fact that b is small is not arbitrary. With this choice the derivative of W will be small with respect to the derivative of V, so the rate of change of W will be slow with respect to the rate of change of V. The dynamic of the potential is quicker than the dynamic of the potassium channel. In fact the neuron usually has two kinds of variables : a *fast* changing variable (the potential V) and a *slowly* changing variable (the potassium conductance).

We need to give some definitions in order to show the conditions for having a repetitive sequence of spikes from the FN model. We are going to give a simple example of these definitions in the course of the proof of the main result concerning the model but we can also refer to the example of the pendulum for understanding and have a simple representation of the definitions we are going to give. The equation of the pendulum of mass m of length l is

$$m\frac{d^2\theta}{dt^2} = -m\frac{g}{l}\sin\theta$$

where θ is the angle between the vertical passing through the point where the pendulum is attached and the inclination of the bar. Let us write the equation as a system introducing the angular velocity v_θ.

$$v_\theta = \frac{d\theta}{dt}.$$

$$\begin{cases} \frac{d\theta}{dt} = v_\theta \\ \frac{dv_\theta}{dt} = -\frac{g}{l}\sin\theta \end{cases}$$

The values of the variables θ, v_θ which make the right-hand side of the equation equal to zero are $v_\theta = 0$, $\theta = 0$. It means physically that the pendulum is in the vertical position with zero angular velocity. This point can be found intersecting the two sets of points where the r.h.s. are zero, these sets of points being the *null clines* of the system

$$v_\theta = 0$$

$$\theta = 0$$

Clearly the point $(0,0)$ is a solution of the equation and it is called *fixed point*. It is a well known fact that if the pendulum is put in this position it remains there. Suppose now that the pendulum starts from the position $(\theta_0, 1)$. Then it will make isochronous oscillations of period $T = 2\pi\sqrt{l/g}$ around the vertical indefinitely with the angle $\theta(t) \in (-\theta_0, \theta_0)$ and $v_\theta \in (-1, 1)$. Thus the motion will always be inside the set $U \equiv (-\theta_0, \theta_0) \times (-1, 1)$. This property is called *stability*. Suppose we know that a friction is added to the pendulum then the equations will be different, $c > 0$ is a friction coefficient.

$$\begin{cases} \frac{d\theta}{dt} = v_\theta \\ \frac{dv_\theta}{dt} = -\frac{g}{l}\sin\theta - cv_\theta \end{cases}$$

Now the fixed point $(0,0)$ remains the same as it is easy to verify but the stability property changes in a stronger condition the *asymptotic stability*. In fact the pendulum will not have isochronous oscillations but oscillations of smaller and smaller amplitudes around the position $\theta = 0$ since the pendulum looses energy, and the motion converges to the point $\theta = 0$, $v_\theta = 0$ which is now an *asymptotically stable fixed point*. All these concepts can be given in more general terms valid for any situation in the following set of definitions. We give further illustration by plotting the solution of the equations for the pendulum in the case of stability (Fig. 3.1) and asymptotic stability (Fig. 3.2). In the pictures the case of small oscillations is displayed.

Definition 3.2. Null-cline.
Consider a system of differential equations for the variables Y_1, Y_2

$$\begin{cases} \frac{dY_1}{dt} = F_1(Y_1, Y_2) \\ \frac{dY_2}{dt} = F_2(Y_1, Y_2) \end{cases}$$

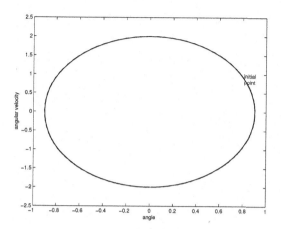

Fig. 3.1 The origin in the phase space (θ, v_θ) of the pendulum is a stable fixed point, the trajectory remains always in the rectangle $U = (-1, 1) \times (-2.5, 2.5)$.

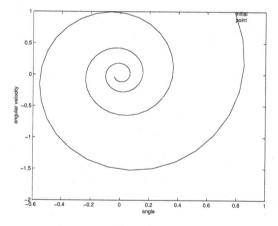

Fig. 3.2 The origin of the phase space of the pendulum becomes an asymptotically stable fixed point if a friction is included in the system. For any initial point the solution goes to the origin.

The null-clines of the system are the curves defined by the equations

$$\begin{cases} F_1(Y_1, Y_2) = 0 \\ F_2(Y_1, Y_2) = 0 \end{cases}$$

Definition 3.3. Fixed point.

Consider a system of differential equations for the variables Y_1, Y_2

$$\begin{cases} \frac{dY_1}{dt} = F_1(Y_1, Y_2) \\ \frac{dY_2}{dt} = F_2(Y_1, Y_2) \end{cases}$$

a point $\overline{Y_1}, \overline{Y_2}$ is a fixed point if the functions $Y_1(t) = \overline{Y_1}$, $Y_2(t) = \overline{Y_2}$ are solutions of the system.

It follows immediately from these two definitions that a point Y_1, Y_2 is a fixed point of the system of differential equations if it is the intersection of two null clines.

Definition 3.4. Stability.

A fixed point $\overline{Y_1}, \overline{Y_2}$ of the system

$$\begin{cases} \frac{dY_1}{dt} = F_1(Y_1, Y_2) \\ \frac{dY_2}{dt} = F_2(Y_1, Y_2) \end{cases}$$

is stable if there exists a neighborhood U of $\overline{Y_1}, \overline{Y_2}$ such that for any initial condition $(Y_1(0), Y_2(0)) \in U$ the solution $(Y_1(t), Y_2(t))$ of the system which satisfies this initial condition remains in U.

Definition 3.5. Asymptotic stability.

A fixed point $\overline{Y_1}, \overline{Y_2}$ of the system

$$\begin{cases} \frac{dY_1}{dt} = F_1(Y_1, Y_2) \\ \frac{dY_2}{dt} = F_2(Y_1, Y_2) \end{cases}$$

is asymptotically stable if there is a neighborhood U of $\overline{Y_1}, \overline{Y_2}$ such that all the solutions starting from U converge, for large t, to the fixed point.

After these generalities we can come back to the FN model. It will be seen in the next section that the fixed points of the model have a very important role as we can understand also from our discussion of the pendulum. First let us analyze the null-clines of the model since the fixed points of the model are the intersection of the null-clines. These are:

$$\begin{cases} V(1-V)(V-a) - W + I = 0 \\ V - \gamma W = 0 \end{cases}$$

Since we choose the variable W to be the dependent variable we can rewrite the system as:

$$\begin{cases} W = V(1-V)(V-a) + I = 0 \\ W = V/\gamma \end{cases}$$

Thus the first null cline is a cubic and the second is a straight line passing through the origin. We note that for $I = 0$ the intersection of the two null-clines is the origin (Fig. 3.3). So for zero input current the fixed point of the FN model is the origin. In the lemma of the next section we show that this point is always asymptotically stable and so there are no spikes because all the oscillations of the system will decrease to zero. For I large enough the intersection of the null-clines changes as shown in Fig. 3.4. In this case there are stable oscillations, i.e. there is the spiking activity as we are going to show in the next section.

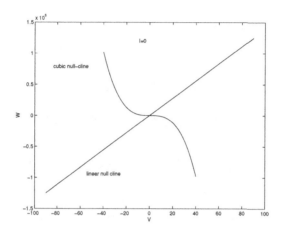

Fig. 3.3   Intersection of the two null-clines of the FN model for $I = 0$. The fixed point is the origin and it is asymptotically stable so there is no spiking activity.

3.3   The generation of spikes and the Hopf bifurcation

In this section we show the results of Lemma 3.1. In practice we will show that the fixed point $(0,0)$ obtained by intersecting the null clines of the FN model, in the case when $I = 0$, like in the case of Fig. 3.3, is always asymptotically stable and so all the oscillations of the $V(t), W(t)$ variables go to zero similarly to what is shown in Fig. 3.2 for the pendulum. This case is not interesting for us because there are no spikes. If I increases then the fixed point moves but it always remains asymptotically stable. We show that there is a value of $I = I_1$ such that the fixed points V_1, W_1 loses then stability property and then a periodic motion appears around it according to a general mathematical mechanism called *Hopf bifurcation* that we will

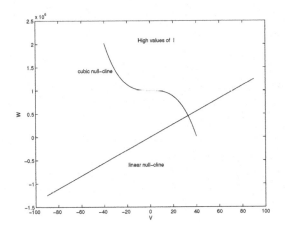

Fig. 3.4   Intersection of the two null-clines of the FN model for high values of I. The fixed point is far from the origin and is no more asymptotically stable.

examine with some detail in the proof of the Lemma 3.1.

Lemma 3.1.

For the case of the Fitzhugh-Nagumo model (FN model) there are two values of the external currents I, I_1, I_2

$$I_i = V_i(V_i^2 - (a + 1)V_i + (a + 1/\gamma)) \tag{3.3}$$

where V_i, W_i are the solutions of the system

$$\begin{cases} V(1 - V)(V - a) - W + I = 0 \\ V - \gamma W = 0 \end{cases}$$

$$V_i = (1 + a) \mp \sqrt{(1 + a)^2 - 3(a + b\gamma)} \tag{3.4}$$

$i = 1, 2$ such that for

$$I < I_1 \quad or \quad I > I_2 \tag{3.5}$$

there are no spikes since the solution of the FN model converges to the point (V_1, W_1), which is an asymptotically stable point and for

$$I_1 < I < I_2 \tag{3.6}$$

there is an infinite sequence of spikes since the solution is a stable periodic
orbit. The transition from the asymptotically stable fixed point (no spikes)
to a stable periodic orbit (infinite sequence of spikes) is through a Hopf
bifurcation.

Proof. We first start to analyze the case $I = 0$. In this case the fixed
point is the origin $(0, 0)$ as it is easy to see from the system

$$\begin{cases} W = V(1 - V)(V - a) \\ W = V/\gamma \end{cases}$$

When studying the stability property of a fixed point a system of non-linear
equation like the FN model for the $I = 0$ case (but the same holds also for
the $I \neq 0$ case)

$$\begin{cases} \frac{dV}{dt} = V(1 - V)(V - a) - W \\ \frac{dW}{dt} = b(V - \gamma W) \end{cases}$$

it is possible to study the solutions in a small neighborhood of the fixed
point. So let us assume that the distances of the points V, W from the origin
are small enough $|V|, |W| \leq \epsilon$ then the non-linear terms appearing in the
r.h.s. of the FN system can be neglected, we do the so-called *linearization*

$$\frac{dV}{dt} = V(1 - V)(V - a) - W = -V^3 + V^2(1 + a) - aV - W \sim -aV - W$$

$$(3.7)$$

$$\frac{dW}{dt} = bV - b\gamma W \qquad (3.8)$$

Let us look for solutions of exponential type which are typical of the linear
system of differential equations. Thus we write, using the vector U for
denoting the solutions of the system:

$$U = \begin{pmatrix} V(t) \\ W(t) \end{pmatrix} = \begin{pmatrix} V_0 \\ W_0 \end{pmatrix} e^{\lambda t}$$

this is equivalent to set

$$\begin{cases} V(t) = V_0 e^{\lambda t} \\ W(t) = W_0 e^{\lambda t} \end{cases}$$

We insert this solution into the linearized system of differential equations:

$$\begin{cases} V_0 \lambda e^{\lambda t} = -a V_0 e^{\lambda t} V - W_0 e^{\lambda t} \\ W_0 \lambda e^{\lambda t} = b V_0 e^{\lambda t} - b\gamma W_0 e^{\lambda t} \end{cases}$$

and eliminate the exponential $e^{\lambda t}$ getting a homogeneous linear system

$$-(a + \lambda)V_0 - W_0 = 0 \tag{3.9}$$

$$bV_0 - (b\gamma + \lambda)W_0 = 0 \tag{3.10}$$

Since the system is homogenous we can get only the value of the ratio W_0/V_0. Actually such a system can be solved, by general theorems on system of linear equations, only if the determinant of the matrix formed by the coefficients of the equation is zero. We give the definition of matrix and some elements of matrix calculus in appendix C. Here we give a simple proof of our statements. The matrix of the coefficients in our case is

$$A = \begin{pmatrix} -a - \lambda & -1 \\ b & -b\gamma - \lambda \end{pmatrix}$$

and its determinant is

$$\mathrm{Det}A = (a + \lambda)(b\gamma + \lambda) + b \tag{3.11}$$

The fact that this determinant must be zero in order to have that the homogeneous system (3.9) to be solvable can be easily understood from the following elementary argument. It is possible to find the ratio W_0/V_0 both from the first and the second equations of the system (3.9) obtaining two different expressions for it:

$$\frac{W_0}{V_0} - -(a + \lambda) \tag{3.12}$$

$$\frac{W_0}{V_0} = \frac{b}{b\gamma + \lambda} \tag{3.13}$$

In order to have a meaningful answer for the ratio, the two expressions must coincide:

$$\frac{W_0}{V_0} = -(a + \lambda) \tag{3.14}$$

$$= \frac{W_0}{V_0} = \frac{b}{b\gamma + \lambda} \tag{3.15}$$

so one gets the equation

$$-(a + \lambda) = \frac{b}{b\gamma + \lambda} \tag{3.16}$$

We rewrite this equation in the form

$$(a + \lambda)(b\gamma + \lambda) + b = 0$$

and so we get the result we wanted to show

$$\text{Det}A = (a + \lambda)(b\gamma + \lambda) + b = 0 \tag{3.17}$$

If we remind that λ is the coefficient appearing in the exponential of the solution of the linearized system of differential equations (3.7):

$$U = \begin{pmatrix} V_0 \\ W_0 \end{pmatrix} e^{\lambda t}$$

we have obtained a remarkable result, which is very general for the system of linear differential equations, that the value of λ must satisfy the equation:

$$Det(A - \lambda I) = (a + \lambda)(b\gamma + \lambda) + b = 0$$

which is the *secular equation* for the matrix A and so λ is one of the *eigenvalues* of the matrix A (see Appendix C for these definitions). The secular equation of a 2×2 matrix is always a simple quadratic equation in the unknown variable λ:

$$\lambda^2 + \lambda(a + b\gamma) + b(1 + \gamma a) = 0 \tag{3.18}$$

the roots λ_1, λ_2 of this equation are the values which we can put in the exponents of the solution from which we started this analysis. In this way we find two solutions of the linearized system:

$$U = \begin{pmatrix} V_0 \\ W_0 \end{pmatrix} e^{\lambda t}$$

one for each value of λ, the values of λ are

$$\lambda_{1,2} = \frac{-(a + b\gamma) \pm \sqrt{(a + b\gamma)^2 - 4b(1 + \gamma a)}}{2}. \tag{3.19}$$

Note that the eigenvalues are both negative since a, b, γ are all positive numbers and so $(a + b\gamma)^2 - 4b(1 + \gamma a) < (a + b\gamma)^2$ from which it follows that $\sqrt{(a + b\gamma)^2 - 4b(1 + \gamma a)} \leq a + b\gamma$ and so both the quantities $-(a + b\gamma \pm \sqrt{(a + b\gamma)^2 - 4b(1 + \gamma a)}$ are negative. Since we have found two different solutions of the system of homogeneous linear differential equations we have that any combination of these two solutions is still a solution of the system so we get that the general solution of our system

$$\frac{dV}{dt} = -aV - W \tag{3.20}$$

$$\frac{dW}{dt} = bV - b\gamma W \tag{3.21}$$

has the form

$$V(t) = V_1 e^{\lambda_1 t} + V_2 e^{\lambda_2 t} \tag{3.22}$$

$$W(t) = W_1 e^{\lambda_1 t} + W_2 e^{\lambda_2 t} \tag{3.23}$$

$$\tag{3.24}$$

Since both the exponents appearing in this formula are negative, all the exponentials will go to zero for t large. We conclude that, if we are in a small neighborhood of the origin where it is possible to approximate the original equations of the FN model with the linearized system (3.20), all the solutions will go to zero with exponential velocity and there will be no oscillations, in other terms no spiking activity can be done by this neuron. The graph of its solution is of the type of Fig. 3.2. The origin is asymptotically stable, it means that the potential V and the W variable of the model will go to zero and the neuron will remain silent. This happens in the case $I = 0$ and so this case is not interesting for us. So we concentrate now on the case $I \neq 0$, for values of I such that the intersection of the null-clines are as in Fig. 3.4 when the fixed point is far from the origin and there is only one intersection of the two null clines. Now we have to find the new fixed point V_c, W_c of the system

$$\frac{dV}{dt} = V(1 - V)(V - a) - W + I \tag{3.25}$$

$$\frac{dW}{dt} = bV - b\gamma W \tag{3.26}$$

let us define for simplicity $f(V) = V(1 - V)(V - a)$, the system for the fixed point is

$$f(V_c) - W_c + I = 0 \tag{3.27}$$

$$V_c - \gamma W_c = 0 \tag{3.28}$$

Substituting the expression $W_c = V_c/\gamma$ from the second equation of the system (3.27) we obtain a single cubic equation

$$f(V_c) - \frac{V_c}{\gamma} + I = 0 \tag{3.29}$$

Now we will study the stability of the new fixed point $P_c \equiv (V_c, W_c)$. Let us introduce the notation $F_1(V, W) = f(V) - W + I$ and $F_2(V, W) = b(V - \gamma W)$. The linearization procedure of the system in the neighborhood of the fixed point is simply obtained by Taylor expanding F_1 and F_2 around P_c:

$$F_1 = F_1(V_c, W_c) + F_{1V}(V - V_c) + F_{1W}(W - W_c) + G_1(V - V_c, W - W_c)$$
$$F_2 = F_2(V_c, W_c) + F_{2V}(V - V_c) + F_{12W}(W - W_c) + G_2(V - V_c, W - W_c)$$

where $F_{1V}, F_{1W}, F_{2V}, F_{12W}$ are the partial derivatives of F_1, F_2 with respect to V and W computed at the point P_c and $G_1(V - V_c, W - W_c), G_2(V - V_c, W - W_c)$ are second order terms that can be neglected since we study the behavior of the system in a small neighborhood of the fixed point P_c. These derivatives are easy to compute.

$$\begin{cases} F_{1V} = f'(V_c) = -3V_c^2 + 2V_c(a + 1) - a \\ F_{1W} = -1 \\ F_{2V} = b \\ F_{2W} = -b\gamma. \end{cases}$$

The linearized FN model around the point P_c is

$$\frac{dV}{dt} = f'(V_c)(V - V_c) - (W - W_c) \tag{3.30}$$

$$\frac{dW}{dt} = b(V - V_c) - b\gamma(W - W_c) \tag{3.31}$$

In order to write the system in a more compact form let us introduce the new variables $\overline{V} = V - V_c$ and $\overline{W} = W - W_c$, so that the system takes the form

$$\frac{d\overline{V}}{dt} = f'(V_c)\overline{V} - \overline{W} \tag{3.32}$$

$$\frac{d\overline{W}}{dt} = b\overline{V} - b\gamma\overline{W} \tag{3.33}$$

Now the system is of the type of the one studied for $I = 0$ but the coefficients are different. Thus we apply the same method, we suppose that the solution is of the form of an exponential

$$U = \left(\frac{\overline{V_0}}{\overline{W_0}}\right) e^{\lambda t}$$

Substituting this solution inside the system and doing the same calculations as in the case $I = 0$ we get the equation

$$\text{Det}(A - \lambda I) = 0 \tag{3.34}$$

Where the matrix A now is

$$A = \begin{pmatrix} f'(V_c) & -1 \\ b & -b\gamma \end{pmatrix}$$

and so

$$A - \lambda I = \begin{pmatrix} f'(V_c) - \lambda & -1 \\ b & -b\gamma - \lambda \end{pmatrix}$$

Applying the definition of the determinant we get the secular equation for the new eigenvalues λ:

$$-(f'(V_c) - \lambda)(b\gamma + \lambda) + b = 0 \tag{3.35}$$

let us write this equation in explicit form:

$$\lambda^2 + (b\gamma - f'(V_c))\lambda + b(1 - \gamma f'(V_c)) = 0$$

The solution is

$$\lambda_{1,2} = \frac{-(b\gamma - f'(V_c)) \pm \sqrt{(b\gamma - f'(V_c))^2 - 4b(1 - \gamma f'(V_c))}}{2}. \tag{3.36}$$

Suppose that $b\gamma - f'(V_c) > 0$ and $0 < 1 - \gamma f'(V_c) < b\gamma - f'(V_c)$ then both the eigenvalues λ_1, λ_2 are negative and the point P_c is asymptotically stable since the solution of the linearized system (3.32) is the combination of the two solutions

$$\overline{V(t)} = V(t) - V_c = \overline{V_1}e^{\lambda_1 t} + \overline{V_2}e^{\lambda_2 t} \qquad (3.37)$$

$$\overline{W(t)} = W(t) - W_c = \overline{W_1}e^{\lambda_1 t} + \overline{W_2}e^{\lambda_2 t} \qquad (3.38)$$

and both exponentials go to zero because of the negative sign of λ_1, λ_2. If $1 - \gamma f'(V_c) > b\gamma - f'(V_c)$ the two eigenvalues are complex conjugate but with a negative real part so the solutions converge to P_c also in this case but making a spiral like in the case of Fig. 3.2. Suppose now that this is the case so $\lambda_1 = \Re\lambda + i\Im\lambda$ and $\lambda_2 = \overline{\lambda_1} = \Re\lambda - i\Im\lambda$, where $\Re\lambda$ is the real part of λ and $\Im\lambda$ is the imaginary part of λ, $\Re\lambda < 0$. If $\Re\lambda = 0$ the point P_c will not be any more asymptotically stable and a periodic motion appears around P_c since the two exponentials generate a periodic function. In fact

$$e^{\lambda_1 t} = e^{(\Re\lambda + i\Im\lambda)t}$$
$$= e^{i\Im\lambda t} = \cos\Im\lambda t + i\sin\Im\lambda t$$
$$e^{\lambda_2 t} = e^{(\Re\lambda - i\Im\lambda)t}$$
$$= e^{-i\Im\lambda t} = \cos\Im\lambda t - i\sin\Im\lambda t$$

Inserting these expressions in the equation for the solutions (3.37) we get that, in the case $\Re\lambda = 0$, the solution is of the form

$$\overline{V(t)} = V(t) - V_c = A\cos\Im\lambda t + B\sin\Im\lambda t \qquad (3.39)$$

$$\overline{W(t)} = W(t) - W_c = C\cos\Im\lambda t + D\sin\Im\lambda t \qquad (3.40)$$

where A, B, C, D are some constants, so it becomes a periodic orbit around the point P_c with period T given by

$$T = \frac{2\pi}{\Im\lambda} \qquad (3.41)$$

Thus an asymptotic stable fixed point disappears and a periodic orbit around it will appear when $\Re\lambda$ changes sign. The change of the type of solution of the model equations is called, in the language of dynamical systems , a *bifurcation* . The transition from an asymptotically stable motion to a periodic motion, of the type described above, is named *Hopf bifurcation*[Hassard et al. (1981)]. The value of the period and the characterization

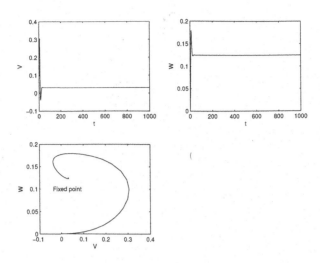

Fig. 3.5 $I < I_1$, $\alpha = -0.3$. The fixed point $P_c \sim (0.3, 0.13)$ is asymptotically stable, no spikes for FN model, the solution converges exponentially fast to P_c.

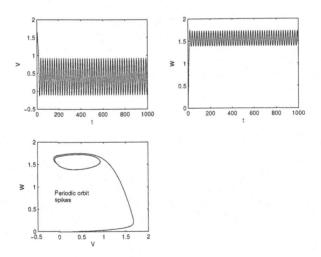

Fig. 3.6 $\alpha = 0.3$, $I_1 < I < I_2$ there is an attractive periodic orbit around the fixed point which is changed $P_c \sim 0.5, 1.5$, the spiking activity is indefinite.

of the bifurcation is the content of the Hopf's theorem. The periodic motion corresponds to an infinite sequence of spikes and is the situation interesting for the neural modelling. Now we have to show that the values of I given in the Lemma are those for which the Hopf's bifurcation takes place. From

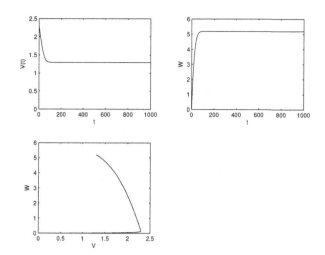

Fig. 3.7 $\alpha = 2$, $I > I_2$ the spiking activity, i.e. the periodic orbit disappeared and again the fixed point $P_c \sim (1.4, 5)$ becomes asymptotically stable.

the discussion made above we have that the condition for having a periodic orbit is that the eigenvalue has zero real part $\Re\lambda = 0$ and that is purely imaginary $\Im\lambda \neq 0$. This happens if

$$b\gamma - f'(V_c) = 0 \tag{3.42}$$

$$1 - \gamma f'(V_c) > 0 \tag{3.43}$$

The second inequality is easy to satisfy if one uses the first equality

$$1 - \gamma f'(V_c) = 1 - \gamma^2 b > 0 \rightarrow \gamma^2 b < 1$$

Let us discuss the first equality

$$f'(V_c) = -3V_c^2 + 2V_c(1 + a) - a = b\gamma \Longrightarrow$$
$$\Longrightarrow 3V_c^2 - 2V_c(1 + a) + a + b\gamma = 0 \Longrightarrow$$
$$\Longrightarrow V_{1,2} = \frac{(1 + a) \pm \sqrt{(1 + a)^2 - 3(a + b\gamma)}}{3}$$

Thus if $V_c = V_1$ or $V_c = V_2$ there will be a Hopf bifurcation. The first bifurcation is the transition from an asymptotically stable motion to a stable periodic motion and the other one is the inverse transition. Let us find now the values of I, I_1 and I_2 where such transitions occur, I being parameter which triggers the transition, I_i are the *critical values*. The connection between I_i and V_i is given by the intersection of the two null-clines or the fixed point condition (3.29):

$$f(V_i) - \frac{V_i}{\gamma} + I_i = 0 \implies$$

$$\implies I_i = V_i(V_i^2 - (a+1)V_i + (a + \frac{1}{\gamma}))$$

Let I_1 be the smaller of the two critical values of I and I_2 the larger. So we get the result of the Lemma because for $I < I_1$ the fixed point P_c is asymptotically stable and there will be no spikes, when $I = I_1$ the fixed point is no more stable and a periodic orbit around it will appear and this happens for all the values $I_1 < I < I_2$. Thus (I_1, I_2) is the region where the spikes of the FN model occur. For $I = I_2$ the periodic orbit disappears and P_c is again asymptotically stable, again the oscillations of the system will be damped. The Lemma is proved. □

In appendix C there is a program which finds the solutions of the FN model using some Matlab functions explained there. We show the results in the three figures Figs 3.5, 3.6, 3.7. The value of the parameters are

$$a = 0.5$$
$$b = 1/10$$
$$\gamma = 1/4$$
$$q = (1+a)^2 - 3(a - b\gamma)$$
$$v_1 = ((1+a) - \sqrt{q})/3$$
$$v_2 = ((1+a) + \sqrt{q})/3$$
$$x = v_1$$
$$I_1 = x(x^2 - x(1+a) + a + 1/\gamma)$$
$$x = v_2$$
$$I_2 = x(x^2 - x(1+a) + a + 1/\gamma)$$
$$\alpha = 0.3$$
$$I = I_1 + \alpha(I_2 - I_1)$$

q is a variable for writing the expressions better and $I_1 < I_2$. α is a parameter for varying I $I < I_1$ or $I_1 < I < I_2$ or $I > I_2$. In the figures the first graph in the upper left is the potential $V(t)$, the second is the recovery current $W(t)$ and the third below is the trajectory $V(t), W(t)$.

3.4 A more realistic model: the Hodgkin-Huxley model (HH model)

The FN model has given an interesting behavior with a minimum input conditions and by means of interesting mathematical mechanism typical of dynamical system analysis, i.e. Hopf's bifurcation. But the full information is still not present in the FN model since the currents used in the equation of the potential V are rather far from the real currents. Furthermore there in the model is not present the phenomena of opening and closing of the ion channels and the distinction between the different currents and channel is rather crude. Even if we got the repetitive sequence of spikes we would like to reproduce another important feature of the neuron dynamics: the *hyperpolarized after potential* (hap). This means that when the neuron emits the spike and starts to hyperpolarization it comes back to values which are less than the initial one. The model of Hodgkin-Huxley (HH model) contains all this phenomenology and has been introduced by Hodgkin and Huxley for reproducing the spike activity of the axon of the squid and it is the first model with the real phenomenology of the neuron, it appeared in a sequence of papers in the fifties [Hodgkin et al. (1952a)], [Hodgkin et al. (1952b)], [Hodgkin et al. (1952c)], [Hodgkin et al. (1952d)] and Hodgkin and Huxley won the Nobel prize in biology for it. Another important feature of the model is that one gets the *refractory period*: after the emission of a spike the neuron is quiescent for some time, during this time interval it does not receive signal from outside. The existence of the hap is a good explanation of this behavior. The important effect which allows the creation of the spike is the change of *permeability* of the selective ion channels during the spike generation. The permeability of a ion channel is a quantity which measures the facility of the ions to go through the channel and is connected with the open or close configuration of the protein which constitute the channel. The depolarization is connected with flowing inside the neuron of Na^+ due to the increased permeability P_{Na^+} of the sodium channel created by the neurotransmitters, the flow is from outside to inside because the Na^+ concentration is higher outside the neuron than inside as one can see from Table 3.1 which lists the concentrations of ions for two kinds of neurons. So we can say in words that the depolarization is caused by the opening of the sodium channel and the inflow of the sodium ions, then the sodium channels close and the potassium channel opens, always with the increase of its permeability P_{K^+}, and then the potassium ions go out and the membrane potential depolarizes. The exit of K^+ ions is again due to the difference of potassium concentrations between outside and inside the neuron shown in Table 3.1.

Table 3.1   Ion concentration in two different types of neurons.

neuron	ion	inside concentration (mmole)	outside concentration (mmole)
squid axon	K^+	410	10
squid axon	N_a^+	49	460
squid axon	Cl^{--}	40	540
cat spinal neuron	K^+	150	5.5
cat spinal neuron	N_a^+	15	150
cat spinal neuron	Cl^{--}	9	125

The change of ion concentration for each spike is compensated by a equilibration mechanism such that exceeding ions (Na^+) are sent outside the neuron and the diminished ions (K^+) are increased letting some K^+ to come in from outside. This important mechanism is called the *ATP*-pump and we do not describe it in detail recommending the reader to find it on some book of neurophysiology or biochemistry. The famous formula of Goldman-Hodgkin-Katz ([Goldman (1943)], [Hodgkin et al. (1949)]) connects the value of the membrane potential with the permeability and ion concentration is the basis of the interpretation of the spike generation we have given.

$$V = \frac{RT}{F} \log \left[\frac{P_K[K]_o + P_{Na}[Na]_o + P_{Cl}[Cl]_i}{P_K[K]_i + P_{Na}[Na]_i + P_{Cl}[Cl]_o} \right] \qquad (3.44)$$

Where $[K]_{i,o}, [Na]_{i,o}, [Cl]_{i,o}$ are respectively the ion concentrations inside and outside the neuron, and P_X the permeability of the channels of the ion X and T is the temperature and R the gas constant. This formula is an equilibrium formula found with thermodynamic and electrostatic considerations that is why these constants appear. According to the interpretation given above the permeabilities change during a spike and also they satisfy different inequalities, so during the depolarization $P_{Na} > P_K, P_{Cl}$, during the hyperpolarization the permeability of the potassium P_K will be maximum, etc. In the HH model the opening and closure of the channels is modelled through the channel parameter which is a function of time and satisfies a differential equation. So in this model the number of variables increases with respect to the FN model: there is the potential V and three channel parameters m, h, n which we discuss below. For each of these variables there is a corresponding differential equation. The main element of the HH model is the equation for the potential which is the usual equation with the input currents $J_i = g_i(V - V_i)$ with g_i being the conductances of the ions considered in the model. The first equation is then

$$C\frac{dV}{dt} = g_{Na}(V - V_a) + g_K(V - V_K) + g_{Cl}(V - V_{Cl}) \tag{3.45}$$

where we suppose that the all the points of the neuron membrane have the same potential. The HH model gives a useful form of the conductances g_{Na}, g_K, g_{Cl} that we are going to describe. The conductances g_{Na}, g_K are not simple constant but depend on the potential, while g_{Cl} is a simple constant. The dependence on the potential is through the activation parameters which are defined by

$$g_{Na} = \overline{g_{Na}}m^3h \tag{3.46}$$

$$g_K = \overline{g_K}n^4 \tag{3.47}$$

the variable defining the neuron are thus four V, m, n, h. V and h are fast variables while n and m are slow variables. The parameters m, n, h vary among $(0, 1)$. n is the activation parameter of the potassium channel, m is the activation parameter of the sodium channel and h inactivation variable so m starts from 0 and goes to 1 while h starts from 1 and goes to 0 since it is an inactivation parameter. The parameters m, n, h depend on the potential through a linear differential equation with coefficients depending on the potential:

$$\frac{dm}{dt} = \alpha_m(1 - m) - \beta_m m \tag{3.48}$$

$$\frac{dh}{dt} = \alpha_h(1 - h) - \beta_h h \tag{3.49}$$

$$\frac{dn}{dt} = \alpha_n(1 - n) - \beta_n n \tag{3.50}$$

If the potential of the neuron is kept constant by injection of a suitable current (*patch-clamp experiment*) then the behavior of the parameters as a function of time is

$$m(t) = m_\infty + (m_0 - m_\infty)e^{-t/\tau_m} \tag{3.51}$$

$$h(t) = h_\infty + (h_0 - h_\infty)e^{-t/\tau_h} \tag{3.52}$$

$$n(t) = n_\infty + (n_0 - n_\infty)e^{-t/\tau_n} \tag{3.53}$$

where m_0, h_0, n_0 are initial values at time $t = 0$ for the parameters and

$$m_\infty = \alpha_m/(\alpha_m + \beta_m)$$
$$h_\infty = \alpha_h/(\alpha_h + \beta_h)$$
$$n_\infty = \alpha_n/(\alpha_n + \beta_n)$$
$$\tau_m = 1/(\alpha_m + \beta_m)$$
$$\tau_h = 1/(\alpha_h + \beta_h)$$
$$\tau_n = 1/(\alpha_n + \beta_n)$$

the dependence on the potential of the α and β parameters are

$$\alpha_m(V) = \frac{25 - V}{10[e^{(25-V)/10} - 1]} \tag{3.54}$$

$$\beta_m(V) = 4e^{-V/18} \tag{3.55}$$

$$\alpha_h(V) = \frac{7}{100}e^{-V/20} \tag{3.56}$$

$$\beta_h(V) = \frac{1}{e^{(30-V)/10} - 1} \tag{3.57}$$

$$\alpha_n(V) = \frac{10 - V}{100[e^{(10-V)/10} - 1]} \tag{3.58}$$

$$\beta_n(V) = \frac{1}{8}e^{-V/80} \tag{3.59}$$

All these functions have been found from the experiments. Finally the system of the equations of HH model is

$$C\frac{dV}{dt} = \overline{g_{Na}}m^3h(V - V_a) + \overline{g_K}n^4(V - V_K) + g_{Cl}(V - V_{Cl}) + I \tag{3.60}$$

$$\frac{dm}{dt} = \alpha_m(1 - m) - \beta_m m \tag{3.61}$$

$$\frac{dh}{dt} = \alpha_h(1 - h) - \beta_h h \tag{3.62}$$

$$\frac{dn}{dt} = \alpha_n(1 - n) - \beta_n n \tag{3.63}$$

The study of the bifurcations of this model is too complicated for this elementary introduction and so we give only the graphs with the output, from them it is clear that there is a transition like the FN model. For $I = 1$ we obtain asymptotic fixed point and so there is no spiking (Fig. 3.8), for $I = 10$ there is a spiking activity (Fig. 3.9), actually the phase portrait is more complicated that the one we show here. It is interesting to note in this spiking case that there is also an hap. The programs are given in appendix C.

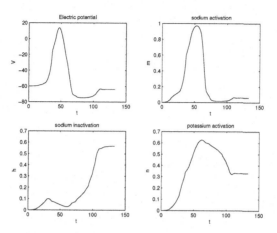

Fig. 3.8 The potential and the activation parameters of the HH model for $I = 1$, no spiking.

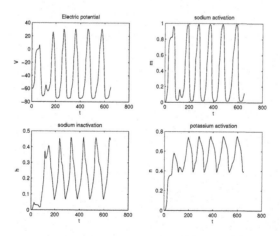

Fig. 3.9 The potential and the activation parameters of the HH model for $I = 10$ there is spiking.

Chapter 4

Definition and simulation of the main random variables

4.1 Introduction

In the previous chapters we have introduced the deterministic model of neurons, the Integrate & Fire model and the FitzHugh-Nagumo model. We gave some solutions of them in the simplest cases. The input current to a neuron from the other neurons has been always schematized as a constant or a function which assumes different simple (linear) forms in different time intervals. But this assumption is too different from the reality. The signals the neurons send to each other are emitted randomly since the neuronal circuit is very complicated, for example in the cerebral cortex each neuron is connected in the average to 60000 neurons and it is difficult that all the neurons fire at the same time with the same intensity. Thus we need to introduce a probabilistic description in order to model more accurately the real behavior of the neurons. We will assume that the input is a stochastic process of a certain type, the Poisson process and simulate it with some Matlab routines. In order to do this we have to explain, since this is a book for beginners, the concept of random variable and how one can simulate a given random variable with a computer. Of course it will also be explained what the words "simulate a random variable" mean with theoretical and practical explanations. The Matlab programs are given explicitly in appendix D. Some introduction to the main probability concepts is also given in this appendix in the form of some simple exercises. In this chapter, after the explanations of the main methods we will proceed by describing the random variables more connected with the neural dynamics and how to simulate them using a computer.

4.2   General definitions

Consider the throwing of a dice with head and cross. The simplest random variable, the *head and cross* is obtained with a launch of a dice enough homogeneous to have equal chances to show head or cross without any preference. If we associate to the event *getting a head* with the value $+1$ and the event *getting a cross* with the value -1 we have defined a random variable X which takes with equal probability the values ± 1:

$$X = \begin{cases} +1 \text{ if head comes out} \\ -1 \text{ if cross comes out} \end{cases} \tag{4.1}$$

The probability of the event $X = 1$ is given, according to the well known definition of probability, as the ratio of the number of positive cases divided by the total number of cases

$$\text{Prob}(X = 1) = \frac{\# \text{ of positive cases}}{\# \text{ number of possible cases}} = \frac{1}{2}. \tag{4.2}$$

The other way of definition of the probability of an event is by means of the *frequency*. This approach is the one directly connected with the simulations and we repeat it here for introducing the idea of what is a simulation of a random variable or a random process. In order to construct the probability from the frequency of events we need to have a sequence of independent trials of the same event and then the probability of the event $A \equiv (X = 1)$ will be given by the frequency of this event $\nu_N(X = 1)$ defined as

$$\nu_N(X = 1) = \frac{\# \text{ of trials with } X = 1}{\# \text{ number of all trials } = N} \tag{4.3}$$

for large values of N, i.e. for $N \to \infty$. In this case $\nu_N(X = 1) \to \frac{1}{2}$, for $N \to \infty$. In words we have that, if the sequence is long enough, the number of *heads* is equal to the number of *crosses*. Simulating a random variable with a given probability distribution means to construct a sequence of independent trials such that the frequency of the events converges to their probability. If one does not use a computer in order to simulate a simple random variable which takes values ± 1 with probability $1/2, 1/2$ it is necessary to throw a coin for example $N = 1000$ times, and to make a record of the outputs of the tosses, so it would take some time. The computer simulates the $X = \pm 1$ random variable by giving, in a much shorter time, a sequence of independent variables of this kind. The condition of independence of the launches means that, if we call the output of the first toss X_1 and of the second toss X_2,

$$\text{Prob}(X_1 = 1, X_2 = -1) = \text{Prob}(X_1 = 1)\text{Prob}(X_2 = -1). \tag{4.4}$$

See appendix D for a more complete presentation of the concept of independence of random events. But the simulation of a random variable by means of a computer gives more information than just the sequence of random variables. For example one can compute the *mean* of the random variable and also the *variance*. These are important definitions which will be used throughout the rest of the book. Let us start with the simple case of a random variable y taking a finite number of values $A \equiv (a_1, a_2, \ldots, a_N)$, in our case $A = (-1, 1)$.

Definition 4.1.

Let $\text{Prob}(X = a)$ be the *probability distribution*, i.e. the probability of the event $X = a$, we define the average of X, and indicate it with EX, the sum

$$EX = \sum_{a \in A} a\text{Prob}(X = a). \tag{4.5}$$

The variance of X is indicated by DX and is

$$DX = E(X - EX)^2 = \sum_{a \in A} (a - EX)^2 \text{Prob}(X = a) \tag{4.6}$$

An evident property follows from the definition of probability

$$\sum_{a \in A} \text{Prob}(X = a) = 1. \tag{4.7}$$

Since the sum over a gives the probability of all the possible events which is equal to 1 by definition, one can use the sequence of the *samples* $X_1, \ldots, X_N$ for computing EX and DX even if the probability distribution $\text{Prob}(X = a)$ is not a priori known as it happens in the majority of the cases. In fact the main content of the statistical estimates is that

$$EX \sim \frac{1}{N} \sum_{i=1}^{N} X_i \tag{4.8}$$

$$DX \sim \frac{1}{N} \sum_{i=1}^{N} (X_i - EX)^2 \qquad (4.9)$$

In the case of the random variable $X = \pm 1$ the average is simple to compute:

$$EX = (1)\frac{1}{2} + (-1)\frac{1}{2} = 0$$

and the variance also

$$DX = EX^2 = (1)^2 \frac{1}{2} + (-1)^2 \frac{1}{2} = \frac{1}{2} + \frac{1}{2} = 1$$

Thus the numbers of $+1$ and -1 in the sample sequence must be approximately equal: the general meaning of the example shown is that from the sample sequence of a random variable, it is possible to estimate its average and its variance directly without explicit knowledge of $\mathrm{Prob}(X)$, on the contrary if these quantities are known then these approximate relationships are used for testing the efficiency of the simulation. The definitions given up to now hold for random variables which take values ± 1 or values in a discrete set while very often one has to deal with variables which take values in a continuous set like for example a variable taking values in the interval $(0, 1)$ or on all the real line $(-\infty, +\infty)$. In this case the concept of probability must be generalized introducing the distribution function (DF). The definition of the DF of a random variable X is:

Definition 4.2. *Distribution Function (DF)*

Given a continuous random variable X, the Distribution Function (DF) $F(x)$ of X is:

$$F(x) = \mathrm{Prob}(X \leq x) \qquad (4.10)$$

A general property of the DFs is that they are monotone increasing in x: $F(x) \leq F(x')$ if $x \leq x'$. This property is evident: the probability that $X \leq x$ increases when x increases because the event $X \leq x'$ contains the event $X \leq x$. Furthermore if $F(x) = 1$ for a given x then it is equal to 1 for any $x' \geq x$. The same holds for the values $F(x) = 0$. If for some x_0 $F(x_0) = 0$, then for any $x \leq x_0$ $F(x) = 0$. An important concept to use when dealing with continuous random variables is the probability density.

Definition 4.3. *Probability density*

Given a continuous random variable X the probability density $f(x)$ is:

$$f(x) = \frac{dF(x)}{dx}. \tag{4.11}$$

The probability density has a simple and interesting interpretation in terms of elementary probabilities. Let us approximate the derivative with the ratio of the increments

$$f(x) = \frac{dF(x)}{dx} \sim \frac{F(x + \Delta) - F(x)}{\Delta}$$
$$= \frac{\text{Prob}(X \leq x + \Delta) - \text{Prob}(X \leq x)}{\Delta} \tag{4.12}$$

where we have used the definition of the distribution function of X, then we remark that $\text{Prob}(X \leq x + \Delta) - \text{Prob}(X \leq x)$ is the probability of X to belong to the interval $(x, x + \Delta)$:

$$\text{Prob}(X \leq x + \Delta) - \text{Prob}(X \leq x) = \text{Prob}(x \leq X \leq x + \Delta)$$

so we get that

$$\text{Prob}(x \leq X \leq x + \Delta) = f(x)\Delta \tag{4.13}$$

this property accounts for the name *density* given to $f(x)$. The idea is that, studying the frequency of the random variable with DF $F(x)$, it is necessary to construct a sample of N values of it. Then the number of points of the sample belonging to the interval $(x, x + \Delta)$ divided by the length of the interval Δ is a density according to the usual sense of this word and is at the same time an approximation of the probability density. A straightforward consequence of this definition is that

$$\text{Prob}(X \in (a, b)) = \int_a^b f(x)dx$$

We explain in these sections how to construct the probability distribution of a given random variable from the samples generated by means of Matlab subroutines or from data samples. The programs which realize the constructions explained in this section and produce the graphs shown in the figures are explained in detail in appendix D. Suppose that a sample $X_1, \ldots, X_N$ of the random variable $X \in (a, b)$ has been constructed using these Matlab routines. $\nu(x, x + \Delta)$, the frequency of the event $X_i \in (x, x + \Delta)$, is given by

$$\nu(x, x + \Delta) = \frac{\#\{k|X_k \in (x, x + \Delta)\}}{N} \qquad (4.14)$$

The numerators of this fraction are the *populations* generated by the sequence. The interval (a, b) is partitioned in M intervals of amplitude Δ, $M = \lceil \frac{b-a}{\Delta} \rceil$. The values of the sample $X_1, \ldots, X_N$ will divide themselves among these intervals and the population is the number of variables in the sample belonging to some interval of the partition. So for each interval $((i-1)\Delta, i\Delta)$ there is an integer m_i of points of the sample belonging to it:

$$m_i = \#\{k|X_k \in ((i-1)\Delta, i\Delta)\}, \quad k = 1, \ldots, N \qquad (4.15)$$

The populations are the set of integers m_i, $i = 1, \ldots, M$. The frequency ν_i relative to the interval $((i-1)\Delta, i\Delta)$ is

$$\nu_i = \frac{m_i}{N} \qquad (4.16)$$

We can summarize the above concepts in the following definition.

Definition 4.4. *Hystogram*
Consider a random variable $X \in (a, b)$ with DF $F(x) = \text{Prob}(X \le x)$ and density $f(x) = \frac{dF(x)}{dx}$. Divide the interval (a, b) in M intervals of equal length $\Delta = \frac{b-a}{M}$, $I_1 = (a, a+\Delta), \ldots, I_M = (a+(M-1)\Delta, b)$. Suppose that a sample of N measures (or realizations) of X are given: $X \equiv (X_1, \ldots, X_N)$ then the histogram of X is the set of frequencies:

$$\nu_n = \frac{\#\{k|X_k \in (a + (n-1)\Delta, a + n\Delta)\}}{N} \qquad (4.17)$$

for $n - 1, \ldots, M$.

As we already mentioned the set of frequencies ν_n converge, for $N \to \infty$ to the probability. This remark allows us to give a simple proof of the relations (4.8) and (4.9) as we shall see in the following sections.

4.3   Uniformly distributed random variable

We show as an example of the definitions of the previous section the variable X of the type $U(0, 1)$, i.e. a uniformly distributed in the interval $(0, 1)$

random variable. This variable is used by us many times for constructing other types of random variables, thus it is very important for our scopes.

Definition 4.5. $U(0,1)$ *random variable*

The $U(0,1)$ random variable X is a uniformly distributed in $(0,1)$ random variable. This means that for any interval $(a,b) \subset (0,1)$ the probability $\mathrm{Prob}(X \in (a,b)) = b - a$.

So the DF of this r.v. is $F(x) = x$, in fact $\mathrm{Prob}(X \leq x) = \mathrm{Prob}(0 \leq X \leq x) = x - 0 = x$. Note that $F(0) = 0$ and $F(1) = 1$ and that $0 \leq F(x) \leq 1$. We show in Fig. 4.1 the DF of a $U(0,1)$ random variable. The probability density of the $U(0,1)$ r.v. is 1 because $F'(x) = 1$.

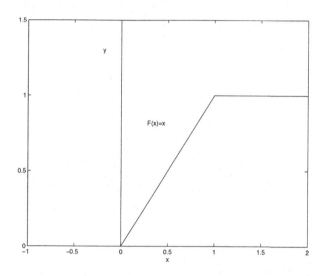

Fig. 4.1   Distribution function for the $U(0,1)$ r.v.

The populations generated by 500 $U(0,1)$ random variables are shown in Fig. 4.2. The interval $\Delta = 0.1$ is constructed by the subroutine.

In Fig. 4.3 a more complete set of graphs is given. The first graph on the left of the uppermost level is the set of 500 $U(0,1)$ variables extracted by means of the random subroutine, then on its right the populations are shown and on the left of the second row there are *frequencies*, i.e. the normalized populations. The last one is the DF extracted from the data evaluated by the Matlab program.

Let us now prove a $U(0,1)$ random variable X and a sample of it $X \equiv (X_1, \ldots, X_N)$ the relations (4.8) and (4.9). Such relations hold in general for

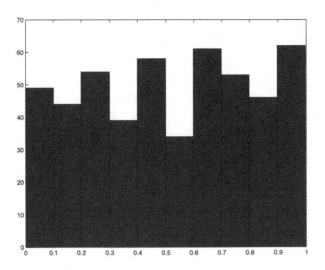

Fig. 4.2   Populations of $U(0, 1)$ random variable.

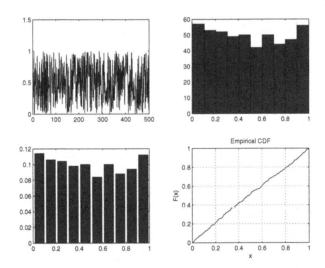

Fig. 4.3   The first picture on the left top corner is the set of 500 $N(0, 1)$ random variables, the one on the right is the populations in the intervals, the first on the left of the second row is the graph of the frequencies, the last on the right is the empirical DF.

any random variable with finite mean and dispersion but we prefer to give the proof for this particular case and show afterwards the generalization to the other cases we are going to consider.

Lemma 4.1.

For a $U(0,1)$ random variable the relations

$$EX \sim \frac{1}{N} \sum_{i=1}^{N} X_i \tag{4.18}$$

$$DX \sim \frac{1}{N} \sum_{i=1}^{N} (X_i - EX)^2 \tag{4.19}$$

hold asymptotically for N large and $EX = 1/2$, $DX = 1/12$.

Proof.

We use the following set of equalities:

$$
\begin{cases}
EX \sim \frac{1}{N} \sum_{i=1}^{N} X_i = \sum_{n=1}^{M} n\Delta \frac{\{\#i | X_i \in ((n-1)\Delta, n\Delta)\}}{N} \\
= \sum_{n=1}^{M} n\Delta \mathrm{Prob}(X \in ((n-1)\Delta, n\Delta)) \\
= \sum_{n=1}^{M} n\Delta f_n \Delta \\
= \sum_{n=1}^{M} x_n f_n \Delta \to_{\Delta \to 0} \\
\to \int_0^1 x \, dx = \\
= 1/2.
\end{cases}
$$

The disjoint intervals $I_n = ((n-1)\Delta, n\Delta)$, $n = 1, \ldots, M$, constitute a partition of the interval $(0,1)$, $\cup_{n=1}^{M} I_n = (0,1)$. M is the number of sets of the partition $M = [\frac{1}{\Delta}]$, where $[x]$ is the integer part of x, i.e. the largest integer less than or equal to x, for example for $x = 1.03$, $[x] = 1$. Let us comment on these formulas.

(a) In the first line the sum over i is done by putting a partition of the interval $(0,1)$ in the intervals $I_n = ((n-1)\Delta, n\Delta)$, where Δ is small enough, and setting all the X_i belonging to the interval I_n equal to its extreme left $n\Delta$ and then multiplying this term by the number $\{\#i | X_i \in ((n-1)\Delta, n\Delta)\}$ of variables of the sample belonging to this interval. In this way an approximation is done which is of the order Δ that can be neglected since Δ is very small and will go to zero at the end of this procedure. The number appearing in the sum $\frac{\{\#i | X_i \in ((n-1)\Delta, n\Delta)\}}{N}$ is thus the frequency of the sample X in the interval I_n.

(b) For $N \to \infty$ the frequency becomes the probability of the variable X to stay in the interval I_n, so this probability appears in the formula replacing the frequency.

(c) We use the formula (4.13) connecting the probability for a variable X to belong to the interval $((n-1)\Delta, n\Delta)$ with the frequency.

(d) We set $x_n = n\Delta$ according to the convention used in the first point.

(e) We notice at this point that the last sum is nothing else but the approximation of the Riemann sums to the integral of the probability density in the interval $(0,1)$. Since the probability density of the $U(0,1)$ random variable is 1, as we have already remarked, letting $\Delta \to 0$ one gets the integral $\int_0^1 dx f(x) x = \int_0^1 x dx$.

(f) The last point is the value of the integral $\int_0^1 x dx = x^2/2|_0^1 = 1/2 - 0 = 1/2$

The same sequence of arguments can be repeated with the evaluation of the dispersion of the random variable X

$$DX \sim \frac{1}{N} \sum_{i=1}^{N} (X_i - EX)^2 \qquad (4.20)$$

the only difference is that instead of having the factor x_n or $n\Delta$ in the sums we have now the term $(n\Delta - EX)^2$ or $(x_n - \frac{1}{2})^2$. The rest of the arguments remains the same.

$$\begin{cases} DX = E(X - EX)^2 \sim \frac{1}{N} \sum_{i=1}^{N} (X_i - EX)^2 \\ = \sum_{n=1}^{M} (n\Delta - EX)^2 \frac{\{\#i | (X_i \in ((n-1)\Delta, n\Delta)\}}{N} \\ = \sum_{n=1}^{M} (n\Delta - EX)^2 \text{Prob}(X \in ((n-1)\Delta, n\Delta)) \\ = \sum_{n=1}^{M} (n\Delta - EX)^2 f_n \Delta \\ = \sum_{n=1}^{M} (x_n - EX)^2 f_n \Delta \to_{\Delta \to 0} \\ \to \int_0^1 (x - 1/2)^2 dx \\ = \frac{(x-1/2)^3}{3} |_0^1 = \frac{(1-1/2)^3}{3} - \frac{(0-1/2)^3}{3} \\ = \frac{1}{24} - (-\frac{1}{24}) - \frac{1}{12} \end{cases} \qquad \square$$

4.4 Exponentially distributed random variables

The exponentially distributed random variables appear in many situations in particular in neurobiology, it is the probability distribution of the time intervals between the arrivals of the spikes as we will see in the next chapter thus we dedicate a section to them.

Definition 4.6. *Exponentially distributed random variable*

The random variable $X \in (0, \infty)$ is exponentially distributed with parameter λ if the DF $F(x)$ is given by

$$F(x) = \text{Prob}(X \leq x) = 1 - e^{-\lambda x} \qquad (4.21)$$

The probability density of this distribution is

$$f(x) = \frac{d}{dx}\text{Prob}(X \leq x) = \frac{d}{dx}(1 - e^{-\lambda x}) = \lambda e^{-\lambda x} \qquad (4.22)$$

The graph of the DF of the exponential distribution is shown in Fig. 4.4 in the case $\lambda = 5$.

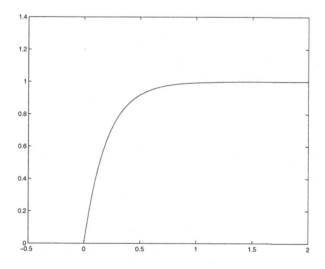

Fig. 4.4   Graph of the DF of the exponential r.v. in the case $\lambda = 5$.

We can use the following Lemma for generating the exponential random variables.

Lemma 4.2.
If X is a $U(0,1)$ r.v. then the variable

$$z = -\frac{1}{\lambda}\log(1 - X) \qquad (4.23)$$

is exponentially distributed with parameter λ.

Proof.

We need to show that

$$\text{Prob}(z \leq A) = 1 - e^{-\lambda A}$$

Let us start from the definition

$$\text{Prob}(-\frac{1}{\lambda}\log(1 - X) \leq A) = \text{Prob}(-\log(1 - X) \leq \lambda A)$$

$$= \text{Prob}(\log(1 - X) \geq -\lambda A) = \text{Prob}(1 - X \geq e^{-\lambda A})$$

$$= \text{Prob}(1 - e^{-\lambda A} \geq X) = 1 - e^{-\lambda A}$$

where in the last line we have applied the property of the $U(0, 1)$ random variables $\text{Prob}(X \leq B) = B$ and all the previous equalities come from the fact that all the sets appearing in the transformations are equivalent and so have the same probability. □

For the exponentially distributed random variables the same lemma as for the $U(0, 1)$ variables holds with the difference that one must start from the interval $(0, M)$ make the same arguments as for the interval $(0, 1)$ and then send $M \to \infty$ in order to get the integral over the infinite interval $(0, \infty)$. We repeat the argument only for completeness. Let $X_1, \ldots, X_N$ be a sample of exponentially distributed random variables with parameter λ. Then

Lemma 4.3.

For an exponential random variable with parameter λ the relations

$$EX \sim \frac{1}{N}\sum_{i=1}^{N} X_i \tag{4.24}$$

$$DX \sim \frac{1}{N}\sum_{i=1}^{N}(X_i - EX)^2 \tag{4.25}$$

hold asymptotically for N large and $EX = \frac{1}{\lambda}$, $DX = \frac{1}{\lambda^2}$.

Proof.

$$
\begin{cases}
EX \sim \frac{1}{N}\sum_{i=1}^{N} X_i = \sum_{n=1}^{M} n\Delta \frac{\{\#i|X_i\in((n-1)\Delta,n\Delta)\}}{N} \\
= \sum_{n=1}^{M} n\Delta \mathrm{Prob}(X \in ((n-1)\Delta, n\Delta)) \\
= \sum_{n=1}^{M} n\Delta f_n \Delta \\
= \sum_{n=1}^{M} x_n f_n \Delta \to_{\Delta \to 0} \\
\to \int_0^A x\lambda e^{-\lambda x}dx \to_{A\to\infty} \\
= \int_0^\infty x\lambda e^{-\lambda x}dx = \frac{1}{\lambda}
\end{cases}
$$

and the intervals $I_n = ((n-1)\Delta, n\Delta)$ constitute a partition of the interval $(0, A)$. M is the number of intervals of size Δ in which the interval $(0, A)$ is divided $M = [\frac{A}{\Delta}]$. The substitution of the frequency with the probability to belong to the interval $((n-1)\Delta, n\Delta))$ is the same and does not depend on the particular random variable. The substitution of the frequency f_n with $\lambda e^{-\lambda x}$ is due to the fact that we are dealing with the exponential random variable and in the limit $\Delta \to 0$ one gets the integral over $(0, A)$. For getting the full average one needs to send $A \to \infty$. Finally the integral over $(0, \infty)$ is computed by making the integral by parts. The formula for the variance is obtained exactly in the same way. Let us compute it.

$$
E(X - \frac{1}{\lambda})^2 = EX^2 - \frac{2}{\lambda}EX + \frac{1}{\lambda^2}
$$

$$
= \frac{2}{\lambda^2} - \frac{2}{\lambda}\frac{1}{\lambda} + \frac{1}{\lambda^2}
$$

$$
= \frac{1}{\lambda^2}.
$$

EX^2 has been computed using the integral by parts

$$
EX^2 = \int_0^\infty x^2 \lambda e^{-\lambda x}dx = \frac{1}{\lambda^2}.
$$

$\square$

We end this section showing a sample of 1000 exponentially distributed r.v. with parameter $\lambda = 5$. The program is shown in appendix D.

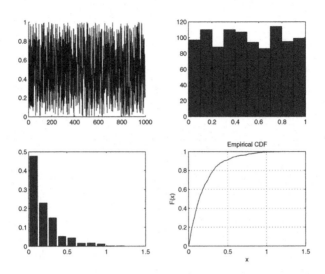

Fig. 4.5 The first graph on the top left shows the values of 1000 exp. r.v. with $\lambda = 5$, on the top right the populations, lower left of the second line displays the frequency, lower right the empirical DF.

4.5   Gaussian random variables

These variables are used in almost any applications where there is some noise to model or to describe the errors of measures using probabilistic methods. In this case N stays for normal, 0 is the average and 1 the variance.

Definition 4.7.
The random variable X is a Gaussian $N(0,1)$ random variable with 0 average and variance 1 if the DF is

$$\text{Prob}(X \leq x) = \int_{-\infty}^{x} e^{-u^2/2} \frac{du}{\sqrt{2\pi}} \tag{4.26}$$

The density $f(x)$ is obtained making the derivative of the DF with respect to x:

$$f(x) = \frac{\text{Prob}(X \leq x)}{dx} = \frac{e^{-x^2/2}}{\sqrt{2\pi}} \tag{4.27}$$

Lemma 4.4.
$EX = 0$ *and* $DX = 1$.

Proof.

That $EX = 0$ follows by inspection of the integral defining the average:

$$EX = \int_{-\infty}^{\infty} u e^{-u^2/2} \frac{du}{\sqrt{2\pi}} \tag{4.28}$$

Since the factor u changes sign while the factor $e^{-u^2/2}$ is even, the result of the integral is zero. $EX^2 = 1$ follows from the definition of the average

$$EX^2 = \int_{-\infty}^{\infty} u^2 e^{-u^2/2} \frac{du}{\sqrt{2\pi}} \tag{4.29}$$

by doing the integral by parts

$$\int_{-\infty}^{\infty} u^2 e^{-u^2/2} \frac{du}{\sqrt{2\pi}} = u \frac{(-e^{-u^2/2})}{\sqrt{2\pi}} |_{-\infty}^{\infty} - \int_{\infty}^{\infty} (-e^{-u^2/2}) \frac{du}{\sqrt{2\pi}} \tag{4.30}$$

$$= \int_{\infty}^{\infty} e^{-u^2/2} \frac{du}{\sqrt{2\pi}} = 1$$

and reminding the normalization property of the Gaussian integrals

$$\text{Prob}(X \leq \infty) = \int_{-\infty}^{\infty} e^{-u^2/2} \frac{du}{\sqrt{2\pi}} = 1. \tag{4.31}$$

$\square$

The convergence of the empirical mean to EX and empirical dispersion to DX follows with the same argument as before and so we will not repeat the argument. We now proceed showing how to simulate a Gaussian r.v. using Matlab routines. We will make use of the erf(x) function.

Definition 4.8.

$$\text{erf}(x) = \frac{2}{\sqrt{\pi}} \int_0^x e^{-u^2} du. \tag{4.32}$$

This function is defined for all $x \in (-\infty, +\infty)$ and is always invertible because it is monotone increasing as one can check also from Fig. 4.6, the program being given in appendix D.

The DF we need to construct is the Gaussian DF

$$\Psi(x) = \text{Prob}(X \leq x) = \int_{-\infty}^{x} e^{-u^2/2} \frac{du}{\sqrt{2\pi}}. \tag{4.33}$$

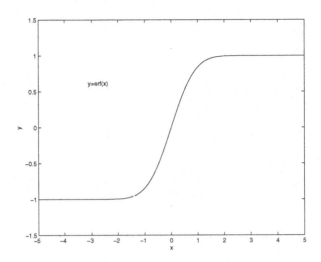

Fig. 4.6   Graph of the function erf(x), it is clear that this function is monotone increasing.

which differs from the erf(x) function that we have just introduced. Some simple transformation allows one to connect the two definitions:

Lemma 4.5.

$$\Psi(x) = (1 + erf(\frac{x}{\sqrt{2}}))/2 \qquad (4.34)$$

We can now give the lemma allowing the construction of the Gaussian $N(0,1)$ random variable.

Lemma 4.6.
If X is a $U(0,1)$ r.v. then

$$Y = \sqrt{2}\,erf^{-1}(2X - 1) \qquad (4.35)$$

is a Gaussian $N(0,1)$ r.v., where $erf^{-1}(z)$ is the inverse function of $erf(z)$.

Proof.
We need to prove that

$$\text{Prob}(Y \leq x) = \int_{-\infty}^{x} e^{-u^2/2} \frac{du}{\sqrt{2\pi}}. \qquad (4.36)$$

We start from

$$\text{Prob}(Y \leq x) = \text{Prob}(\sqrt{2}\,\text{erf}^{-1}(2X - 1) \leq x)$$

$$= \text{Prob}(\text{erf}^{-1}(2X - 1) \leq \frac{x}{\sqrt{2}})$$

$$= \text{Prob}(2X - 1 \leq \text{erf}(\frac{x}{\sqrt{2}}))$$

$$= \text{Prob}(X \leq (1 + \text{erf}(\frac{x}{\sqrt{2}})/2)$$

since X is a $U(0,1)$ r.v. we have, using Lemma 4.5,

$$= (1 + \text{erf}(\frac{x}{\sqrt{2}}))/2 = \int_{-\infty}^{x} e^{-u^2/2} \frac{du}{\sqrt{2\pi}}. \qquad \square$$

We show in Fig. 4.7, as usual, the result of the statistics of a sample of $N = 10000$ Gaussian r.v. obtained with the Matlab program based on this lemma.

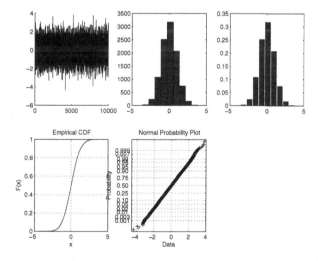

Fig. 4.7 Graph of a Gaussian sample. Following the usual order, there are the values of the random variables, the populations, the histogram, the empirical DF, the last one being a test of normality: if all the points lie on the line $y = x$ then the sample is generated by a Gaussian variable.

We finally describe in the next lemma how to construct a Gaussian random variable with mean μ and variance σ, $N(\mu, \sigma)$ from the $N(0,1)$ Gaussian r.v.

Lemma 4.7.

If X is a $N(0,1)$ r.v. then

$$V = \sigma X + \mu \tag{4.37}$$

is a Gaussian $N(\mu, \sigma)$ r.v., so that

$$\text{Prob}(V \leq x) = \int_{-\infty}^{x} e^{-(u-\mu)^2/(2\sigma^2)} \frac{du}{\sigma\sqrt{2\pi}}. \tag{4.38}$$

Proof.

It is very easy to see that

$$EV = \sigma EX + \mu = \mu \tag{4.39}$$

and

$$E(V - \mu)^2 = E(X\sigma)^2 = \sigma^2 \qquad \square$$

4.6   Poisson random variables

These variables are important for describing the random inputs to a neuron. Consider M different time intervals of unit length. Even if the intervals have the same length the number of spikes arriving at this neuron in each of these intervals is not constant but changes from interval to interval due to the randomness of the neuronal circuit. This number is an example of a Poisson random variable. Other examples are the number of telephone calls arriving at a house in a certain interval of time, the number of alpha particles emitted by a radioactive material in a given interval of time, the number of particles of a gas in a given volume of the space. In this case the interval of time is substituted by the volume. Since for us the number of spikes is the main variable we refer always to this case, although we could speak in more general terms. Proceeding as in the other sections the construction of the empirical DF is done considering the number of arrivals in M different time intervals, but all with unit length, and looking for the frequency of the different values of the number of spikes arriving to the neuron in each of these intervals. Let us call $N(1)$ such a number.

Definition 4.9. *Poisson random variable.*

The variable $N(1)$ is a Poisson random variable if it takes integer values, i.e. $N(1) \in \mathcal{Z}$, $\mathcal{Z}$ being the set of integer numbers, with probability

$$\text{Prob}(N(1) = k) = e^{-\lambda}\frac{\lambda^k}{k!} \tag{4.40}$$

where the parameter λ is the mean value or mean activity of the variable $N(1)$. The mean value of $N(1)$ is

$$EN(1) = \sum_{k=0}^{\infty} k\text{Prob}(N(1) = k) = \sum_{k=0}^{\infty} e^{-\lambda}\frac{\lambda^k}{k!} \tag{4.41}$$

and the variance is

$$DN(1) = \sum_{k=0}^{\infty}(k - EN(1))^2\text{Prob}(N(1) = k) = \sum_{k=0}^{\infty}(k - EN(1))^2 e^{-\lambda}\frac{\lambda^k}{k!} \tag{4.42}$$

If we want to construct a sample of such a variable $N(1)_1, \ldots, N(1)_M$ we have to consider M disjoint time intervals of length 1, $I_1, \ldots, I_M$. $N(1)_1$ is the number of spikes arriving at the neuron in the interval I_1, $N(1)_2$ is the number of spikes arriving at the neuron in the interval I_2 and so on.

Definition 4.10.

The frequency $\nu_M(k)$ of the Poisson variable $N(1)$ is

$$\nu_M(N(1) = k) = \frac{\text{\# of time intervals } I_i \text{ with } N(1) = k \text{ arrivals}}{\text{\# number of all intervals } = M}. \tag{4.43}$$

Our general argument implies that

$$\lim_{M\to\infty} \nu_M(N(1) = k) = \text{Prob}(N(1) = k). \tag{4.44}$$

It is also easy to prove that

Lemma 4.8.

For large values of M the relations hold:

$$EN(1) \sim \frac{1}{M} \sum_{i=1}^{M} N(1)_i \qquad (4.45)$$

$$DN(1) \sim \frac{1}{M} \sum_{i=1}^{M} (N(1)_i - EN(1))^2. \qquad (4.46)$$

Furthermore $EN(1) = \lambda$, $DN(1) = \lambda$.

Proof.

The proof is very similar to the proof of the same lemma for the previous random variables, it is even simpler because in this case the random variable takes integer values, so there is no need to introduce the integral.

$$\begin{cases} EN(1) \sim \frac{1}{M} \sum_{i=1}^{N} N(1)_i = \sum_{n=0}^{\infty} n \frac{\{\#I_i | N(1)_i = n\}}{M} \\ = \sum_{n=0}^{\infty} n \text{Prob}(N(1) = n) \\ = \sum_{n=0}^{\infty} n e^{-\lambda} \frac{\lambda^n}{n!} \\ = \sum_{n=1}^{\infty} e^{-\lambda} \frac{\lambda^n}{(n-1)!} \\ = \lambda e^{-\lambda} \sum_{n=1}^{\infty} \frac{\lambda^{n-1}}{(n-1)!} \\ = \lambda e^{-\lambda} \sum_{l=0}^{\infty} \frac{\lambda^l}{l!} \\ = \lambda \end{cases}$$

The first line is the usual introduction in the sum of the frequency of the values, the second is the substitution of the frequency with the probability. In the fourth line we start the sum from $n = 1$ because the first term is zero and then we simplify $\frac{n}{n!} = \frac{1}{(n-1)!}$, in the fifth line we change index $l = n-1$ so that the sum start again from 0 after having extracted from the sum the factor λ. Finally we use the Taylor expansion of the exponential

$$e^\lambda = \sum_{l=0}^{\infty} \frac{\lambda^l}{l!} \qquad (4.47)$$

for getting the final value.

$$\begin{cases} DN(1) \sim \frac{1}{M} \sum_{i=1}^{N} (N(1)_i - EN(1))^2 \\ = \sum_{n=0}^{\infty} (n - EN(1))^2 \frac{\{\#I_i | N(1)_i = n\}}{M} \\ = \sum_{n=0}^{\infty} (n - EN(1))^2 \text{Prob}(N(1) = n) \\ = \sum_{n=0}^{\infty} (n - \lambda)^2 e^{-\lambda} \frac{\lambda^n}{n!} \\ = \sum_{n=0}^{\infty} (n^2 - 2n\lambda + \lambda^2) e^{-\lambda} \frac{\lambda^n}{n!} \\ = e^{-\lambda} \sum_{n=0}^{\infty} n^2 \frac{\lambda^n}{n!} \\ -2\lambda e^{-\lambda} \sum_{n=0}^{\infty} n \frac{\lambda^n}{n!} + \lambda^2 e^{-\lambda} \sum_{n=0}^{\infty} \frac{\lambda^n}{n!} \end{cases}$$

But the last term of the last line is just λ^2 since

$$e^{-\lambda} \sum_{n=0}^{\infty} \frac{\lambda^n}{n!} = 1$$

while the first expression in this line is $-\lambda^2$ since we have already proved that

$$e^{-\lambda} \sum_{n=0}^{\infty} n \frac{\lambda^n}{n!} = \lambda$$

so we get

$$DN(1) = \lambda e^{-\lambda} \sum_{n=0}^{\infty} n^2 \frac{\lambda^n}{n!} - \lambda^2 \qquad (4.48)$$

The sum in the formula is easily computed

$$e^{-\lambda} \sum_{n=0}^{\infty} n^2 \frac{\lambda^n}{n!} = e^{-\lambda} \sum_{n=1}^{\infty} n \frac{\lambda^n}{(n-1)!} = e^{-\lambda} \lambda \sum_{n=1}^{\infty} n \frac{\lambda^{n-1}}{(n-1)!}$$

$$= e^{-\lambda} \lambda \frac{d}{d\lambda} \sum_{n=1}^{\infty} \frac{\lambda^n}{(n-1)!} = e^{-\lambda} \lambda \frac{d}{d\lambda} \sum_{n=1}^{\infty} \lambda \frac{\lambda^{n-1}}{(n-1)!}$$

$$= e^{-\lambda} \lambda \frac{d}{d\lambda} [\lambda \sum_{n=0}^{\infty} \frac{\lambda^n}{n!}] = e^{-\lambda} \lambda \frac{d}{d\lambda} [\lambda e^{\lambda}]$$

$$= e^{-\lambda} \lambda (e^{\lambda} + \lambda e^{\lambda}) = \lambda^2 + \lambda$$

Inserting this formula in the expression (4.48) we get the result. $\qquad \square$

Note that in the case we are discussing, the $N(1)$ r.v., with the meaning of the number of spikes arriving from the network to a given neuron in the interval of time $(0,1)$, the parameter λ is mean number of spikes arriving at the neuron. This explains the term *activity* used for it. This means that the larger λ the larger the mean number of spikes arrive at the neuron. It is important to stress the fact that we are dealing with the number of arrivals because the frequency of the emission of spikes from a neuron is in general different from the frequency of arriving spikes as we seen in dealing with the *I&F* model where the interspike time was a non-linear function of the input parameters. The only case of coincidence was, under certain conditions, the impulsive spikes.

Now we present, following our general procedure, the lemma describing a way to simulate a Poisson r.v. by means of the $\mathrm{rand}(x)$ subroutine. This time the procedure is different from the others because the variable to be generated takes discrete values while the subroutine $\mathrm{rand}(x)$ takes continuous values in the interval $(0,1)$ so the idea is to generate an event with probability of the type of the Poisson variable.

Lemma 4.9.

If X is a $U(0,1)$ r.v. then a Poisson r.v. $N(1)$ with activity λ is defined by the relation

$$N(1) = k \quad if \quad e^{-\lambda} \sum_{n=0}^{k-1} \frac{\lambda^n}{n!} \leq X \leq e^{-\lambda} \sum_{n=0}^{k} \frac{\lambda^n}{n!} \tag{4.49}$$

Proof.

This lemma says that

$$\mathrm{Prob}(N(1) = k) = \mathrm{Prob}\left(e^{-\lambda} \sum_{n=0}^{k-1} \frac{\lambda^n}{n!} \leq X \leq e^{-\lambda} \sum_{n=0}^{k} \frac{\lambda^n}{n!}\right) \tag{4.50}$$

since the r.v. variable $N(1) = k$ if the event $e^{-\lambda} \sum_{n=0}^{k-1} \frac{\lambda^n}{n!} \leq X \leq e^{-\lambda} \sum_{n=0}^{k} \frac{\lambda^n}{n!}$ takes place. But remember that the $U(0,1)$ r.v. X has been defined just by the condition $\mathrm{Prob}(X \in (a,b)) = b - a$ if $b \geq a$ and if the interval $(a,b) \subset (0,1)$. So if one takes as $a = e^{-\lambda} \sum_{n=0}^{k-1} \frac{\lambda^n}{n!}$ and $b = e^{-\lambda} \sum_{n=0}^{k} \frac{\lambda^n}{n!}$ one gets

$$\mathrm{Prob}(N(1) = k) = \mathrm{Prob}\left(X \in (e^{-\lambda} \sum_{n=0}^{k-1} \frac{\lambda^n}{n!}, e^{-\lambda} \sum_{n=0}^{k} \frac{\lambda^n}{n!})\right)$$

$$= e^{-\lambda} \sum_{n=0}^{k} \frac{\lambda^n}{n!} - e^{-\lambda} \sum_{n=0}^{k-1} \frac{\lambda^n}{n!}$$

$$= e^{-\lambda} \frac{\lambda^k}{k!}$$

and this is the probability distribution of the Poisson variable. We only have to show that

$$b = e^{-\lambda} \sum_{n=0}^{k} \frac{\lambda^n}{n!} < 1.$$

But this is an immediate consequence of the fact that $e^{-\lambda} \sum_{n=0}^{\infty} \frac{\lambda^n}{n!} = 1$. This relationship is the consequence of the Taylor formula we have shown before which can also be interpreted as the probability that $N(1)$ takes any possible integer values which is obviously one. It is evident that $e^{-\lambda} \sum_{n=0}^{k} \frac{\lambda^n}{n!} < e^{-\lambda} \sum_{n=0}^{\infty} \frac{\lambda^n}{n!} = 1$ since there are less terms in the first sum than in the second and by the same argument we get $a < b$. $\qquad \square$

We plot in Fig. 4.8 the graph obtained using the program, described in appendix D, which applies the method of the lemma for constructing respectively the populations, frequencies $f(k)$, of a sample of $N = 1000$ poisson variables corresponding to an activity $\lambda = 2$. The variance computed on the sample approximates the value of λ with an error of 0.050124. $p(k)$, the theoretical Poisson distribution with $\lambda = 2$, and the difference $w(k) = |p(k) - f(k)|$ are also shown in the figure. It can be checked that the errors on the distributions are relatively small.

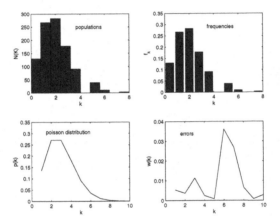

Fig. 4.8   Graph of a Poissonian sample. Following the usual order, there are the values of the populations, the histogram, the theoretical DF and the error done using the frequencies for approximating the real probability distribution.

Chapter 5

Simulation of the neuron dynamics in interaction with a complex network

5.1 Introduction

In the previous section we have given a detailed description of a Poisson r.v. giving as a constant example the r.v. $N(1)$, the number of spikes arriving at a neuron in a time interval of length 1. But this example is very restrictive for the neural dynamics and the evolution of random systems in general since the real object of the study is the knowledge of the variables $N(t)$, the number of events or arrivals happening in the interval $(0, t)$, for any time t. So we are faced to solve the problem of studying a collection of Poisson variables, $N(t), t \in \mathcal{R}$, where $\mathcal{R}$ is the set of real numbers. This object is different from the single variable $N(t)$ for fixed t. In fact it is a *stochastic process*. So we need the definition:

Definition 5.1. *Stochastic process*
   A stochastic process is a set of random variables $X(t)$, $t \in \mathcal{R}$ or $\mathcal{N}$, $X(t) \in \mathcal{N}$ or $\mathcal{R}$. A stochastic process is defined whenever the *joint probability distributions* are given:

$$\text{Prob}(X(t_1) \in A_1, \ldots, X(t_k) \in A_k) = F(t_1, \ldots, t_k, A_1, \ldots, A_k) \qquad (5.1)$$

where $A_1, \ldots, A_k$ are any subsets of $\mathcal{R}$ or $\mathcal{N}$.

It is clear that the description of a stochastic process is more complex, as one can expect, than the one of the single variable. The simplest example is the set of independent random variables.

Definition 5.2. *Process of independent random variables*
   A stochastic process of independent random variables $X(t)$ is such that random variables $X(t)$ are all independent of each other. For each choice of k, $t_1, \ldots, t_k \in \mathcal{R}$ and $A_1, \ldots, A_k \subset \mathcal{R}$ we must have

$$\text{Prob}(X(t_1) \in A_1, \ldots, X(t_k) \in A_k) = \text{Prob}(X(t_1) \in A_1) \ldots \text{Prob}(X(t_k) \in A_k) \tag{5.2}$$

which is equivalent to:

$$F(t_1, \ldots, t_k, A_1, \ldots, A_k) = F(t_1, A_1) \ldots F(t_k, A_k) \tag{5.3}$$

A complete description of a stochastic process can be obtained from the knowledge of the *cylindrical* probability distributions :

$$\text{Prob}(X(t_1) \in I_1, \ldots, X(t_k) \in I_k) = F(t_1, \ldots, t_k, I_1, \ldots, I_k) \tag{5.4}$$

for any possible choice of the integer k, of the times $t_1, \ldots, t_k \in \mathcal{R}$. In this definition we have made a particular choice of the sets of (5.1) $A_1 = I_1, \ldots, A_k = I_k$ where $I_1, \ldots, I_k$ are simple intervals in $\mathcal{R}$ or $\mathcal{N}$. So in order to be able to define a stochastic process we need to define the cylindrical probability distributions (5.4) of $X(t)$. The theorem of Kolmogorov ([Simon (1979)]) guarantees that there is a unique probability description of the stochastic process starting from the knowledge of the cylindrical probability distributions. Another important general property of stochastic processes is stationarity. In words a stochastic process is stationary if the functions $F(t_1, \ldots, t_k, A_1, \ldots, A_k)$ remain the same if all the times $t_1, \ldots, t_k$ are shifted by the same amount τ. For example if we consider the joint probability distribution of $X(t_1)$ and $X(t_2)$ we must have:

$$\text{Prob}(X(t_1) \in A_1, X(t_2) \in A_2) = \text{Prob}(X(t_1 + \tau) \in A_1, X(t_2 + \tau) \in A_2)$$

for any τ. Thus we have the general definition

Definition 5.3. *Stationarity*

A stochastic process $X(t)$ is stationary if its cylindrical probability distributions satisfy the relations:

$$F(t_1, \ldots, t_k, I_1, \ldots, I_k) = F(t_1 + \tau, \ldots, t_k + \tau, I_1 \ldots I_k) \tag{5.5}$$

for each choice of τ, k, $t_1, \ldots, t_k$, $I_1, \ldots, I_k$.

Another consequence of this definition is that if $X(t)$ is stationary then $\text{Prob}(X(t) \in A) = \text{Prob}(X(0) \in A)$ and so the probability distribution of the single variables $X(t)$ does not depend on the time, while the joint probability distribution of two variables depend only on the difference of their times $t_2 - t_1$. It is sufficient to take $k = 2$ and $\tau = -t_1$ in the relation (5.5):

$$\mathrm{Prob}(X(t_1) \in I_1, X(t_2) \in I_2) = \mathrm{Prob}(X(t_1 - t_1) \in I_1, X(t_2 - t_1) \in I_2)$$

$$= F(0, t_2 - t_1, I_1, I_2).$$

From this property we get that the two-point correlation depends only on the time difference:

$$EX(t_1)X(t_2) = EX(0)X(t_2 - t_1).$$

5.2 Definition of a Poisson process

The case which interests us in this chapter is the one when $X(t) = N(t)$. The r.v. variables $N(t)$ are not independent because it is clear that they must satisfy at least one constraint: the monotonicity relation $N(t) < N(t')$ if $t < t'$. In fact if the time increases the number of arrivals increases. So we have to construct a description of a non-independent set of random variables $N(t)$ something which cannot be done by means of the probability distribution of the single variable $N(t)$. We start giving the definition of a Poisson process using, according to the Kolmogorov's theorem, from the definition of the joint probability distributions of the process on cylindrical sets. We can get such information from the following definition:

Definition 5.4. *Poisson process*

A Poisson process is a family of r.v. $N(t) \in \mathcal{N}$, i.e. taking integer values, depending on the continuous time $t \in \mathcal{R}$, stationary in time, with the properties

(1) $N(0) = 0$.
(2) $\mathrm{Prob}(N(t + s) - N(t) = k) = e^{-\lambda s} \frac{\lambda^k}{k!}$.
(3) The increments $N(t_i) - N(t_{i-1})$, $i = 1, \ldots, k$ are independent if the time intervals (t_{i-1}, t_i), $i = 1, \ldots, k$ are disjoints.

The second condition implies that the process is stationary as we will show below, while the third can be used for defining the cylindrical probability distributions as we show in the lemma below. Let us write it explicitly in the case of two disjoint intervals $I_1 \equiv (t_1, t_2)$ and $I_2 \equiv (t_2, t_3)$, $I_1 \cap I_2 = O$. Then the independence condition is given by:

$$\text{Prob}((N(t_3) - N(t_2) = k_2) \cap (N(t_2) - N(t_1) = k_1))$$

$$= \text{Prob}(N(t_3) - N(t_2) = k_2)\text{Prob}(N(t_2) - N(t_1) = k_1).$$

Using such equalities it is possible to define all the probabilities of the cylindrical sets.

Lemma 5.1.

The definition of the Poisson process permits to give all the cylindrical probability distributions.

Proof.

Suppose that $t_1 \leq t_2 \leq \ldots \leq t_k$ with k any arbitrary integer number and also that k integers $a_1, \ldots, a_k$ are given. Then we can compute $\text{Prob}(N(t_1) = a_1, \ldots, N(t_k) = a_k)$ for any choice of t_i, using the second property and third property of the definition:

$$\text{Prob}(N(t_1) = a_1, \ldots, N(t_k) = a_k)$$

$$= \text{Prob}(N(t_1) - N(0) = a_1, \ldots N(t_k) - N(t_{k-1}) = a_k - a_{k-1}).$$

But this probability can be computed using the property of independence since the intervals $t_1 - 0, t_2 - t_1, \ldots, t_k - t_{k-1}$ are disjoints

$$\text{Prob}(N(t_1) - N(0) = a_1, \ldots, N(t_k) - N(t_{k-1}) = a_k - a_{k-1})$$

$$= \text{Prob}(N(t_1) - N(0) = a_1) \ldots \text{Prob}(N(t_k) - N(t_{k-1}) = a_k - a_{k-1})$$

$$= e^{-\lambda t_1} \frac{\lambda^{a_1}}{a_1!} \ldots e^{-\lambda(t_k - t_{k-1})} \frac{\lambda^{a_k - a_{k-1}}}{(a_k - a_{k-1})!}. \qquad \square$$

5.3 The integrate and fire model with Poissonian inputs

We are able now to describe the dynamic of a neuron when it is connected with a complicated network, assuming that the inputs received by the neurons are distributed according to a Poisson process. The definitions given in the previous sections allow us to make such a model. The fact that the inputs coming from the network of neurons to a given neuron of the network are a Poisson process is a strong assumption. It is accepted because it has been found in many simulations done on the basis of this hypothesis that the activity behavior of the real neurons has always been found with a good approximation. An important point is to understand how one can use a Poisson process in the framework of the $I\&F$ model. This is what we are going to explain now. So let us consider an $I\&F$ model with resistance R and capacity C. We have schematized the arrival of a spike to the neuron with a charge q arriving from the other neurons. When the charge q arrives at the neuron its membrane potential $V(t)$ changes immediately according to the simple definition of capacity

$$V(t) = V(t) + \frac{q}{C} \tag{5.6}$$

if $q > 0$ the input is excitatory if $q < 0$ is inhibitory, the neural terminology is EPSP (excitatory post synaptic potential) for excitatory inputs and IPSP (inhibitory post synaptic potential) for the inhibitory inputs. Let us recall the equation for the membrane potential of the $I\&F$ model introduced in chapter 1

$$\begin{cases} \dfrac{dV}{dt} + \dfrac{V}{\tau} = \dfrac{I}{C} \\ V(0) = \overline{V} \\ \text{if} \quad \exists t : V(t) = \theta \rightarrow V(t^+) = 0 \end{cases}$$

we can write it using the increments Δt

$$\frac{V(t + \Delta t) - V(t)}{\Delta t} + \frac{V}{\tau} = \frac{I}{C}$$

which is equivalent to

$$V(t + \Delta t) - V(t) = -\frac{V}{\tau}\Delta t + \frac{I}{C}\Delta t.$$

The quantity $I \times \Delta t$ is the amount of charge ΔQ arriving at the synapses of the neuron in the time interval $(t, t + \Delta t)$:

$$I\Delta t = \Delta Q$$

The charge ΔQ arrives at the neuron because each time a neuron connected with it emits a spike it sends a charge q to it. Let us suppose that this elementary charge is the same for each neuron of the network which sends spikes to our neuron. This assumption can be accepted if one supposes a system of neurons with the same characteristics, i.e. with the same threshold θ and capacity C so that the charge emitted during a spike is always the same amount θC. If $N(t)$ is the number of spikes arriving to our $I\&F$ neuron in the time interval $(0, t)$ we have that

$$\Delta Q = q(N(t + \Delta t) - N(t))$$

because the total charge ΔQ arriving to the neuron in the time interval $(t, t + \Delta t)$ must be equal to the charge q arriving for a single spike $\times$ the number of spikes arriving in the same interval. As we said before it is natural to assume that $N(t)$ is a Poisson process and rewrite the equation for the change of the potential using the Poisson process as input.

$$V(t + \Delta t) - V(t) = -\frac{V}{\tau}\Delta t + \frac{q}{C}(N(t + \Delta t) - N(t))$$

These arguments suggest to introduce the following definition.

Definition 5.5.

An $I\&F$ model with Poisson inputs is defined by the equations:

$$\begin{cases} V(t + \Delta t) - V(t) = -\frac{V}{\tau}\Delta t + a(N(t + \Delta t) - N(t)) \\ V(0) = \overline{V} \\ \text{if} \quad \exists t : V(t) = \theta \rightarrow V(t^+) = 0 \end{cases}$$

where $a = \frac{q}{C}$ and $N(t)$ is a Poisson process with mean λ. If $a > 0$ the input is excitatory and if $a < 0$ the input is inhibitory.

We can interpret the right-hand side of the equation for the potential using the definition of Poisson process. From the definition we know that $N(t + \Delta t) - N(t)$ is a Poisson variable with average $\lambda \Delta t$, so the higher the activity λ of the process the higher the average number of spikes arriving at the neuron and the higher will be the contribution to the increase of the potential. A high λ implies a high spiking frequency but this relation is not so straightforward because the equation for the potential is non-linear and so it is necessary to prove the relationship between λ and the spiking frequency. Qualitatively one can understand the behavior of the solution. The potential $V(t)$ increases of a quantity a when a spike arrives, between the arrival of two spikes the potential decreases as $(V(t_1) + a)e^{-\frac{t-t_1}{\tau}}$ since there is no current. t_1 is the time when the spike arrives and the potential decreases by a factor $e^{-\frac{t-t_1}{\tau}}$ until $t = t_2$ the time of arrival of the next spike. This situation is similar to the one studied in chapter 2 where we studied the evolution of the potential of a neuron which received instantaneous spikes at regular times $t_n = nT$ but there is the big difference that the t_n are no more regular but random times. A consequence of this is that the interspike interval is no more a constant but a random variable described by a probability distribution that can be found from the solution of the equation for the potential. In the next sections we are going to show how one can find the probability distribution of the interspike time in some particular cases and in more general cases using the numerical solution.

5.4 Computation of interspike intervals with Poissonian inputs

We make a simplification of the model equation for arriving at an analytic computation of ET_k, the expectation of the interspike interval T_k. In particular we suppose that the activity of the input Poisson process is very high or, in equivalent terms, that the average of the contribution of the Poisson inputs to the equation for the potential:

$$V(t + \Delta t) - V(t) = -\frac{V(t)}{\tau}\Delta t + a(N(t + \Delta t) - N(t)) \qquad (5.7)$$

is larger than the decay factor. This can be explicitly stated with the inequality

$$aE(N(t + \Delta t) - N(t)) > |\frac{V}{\tau}\Delta t| \qquad (5.8)$$

since $E(N(t + \Delta t) - N(t)) = \lambda \Delta t$ by the definition of the Poisson process we have that this inequality is the same as

$$a\lambda > |\frac{V}{\tau}| \qquad (5.9)$$

so if the time decay τ of the neuron is large or λ (the number of spikes arriving at the unit time) is large the term $\frac{V}{\tau}$ can be neglected from the equation and we remain with the simpler equation

$$V(t + \Delta t) - V(t) = a(N(t + \Delta t) - N(t)) \qquad (5.10)$$

which can be written in the simplest form

$$V(t) = aN(t). \tag{5.11}$$

We can resume the previous discussion and give the results for this model in the following lemma.

Lemma 5.2. *Consider the I&F model with Poissonian inputs:*

$$\begin{cases} V(t + \Delta t) - V(t) = -\frac{V}{\tau}\Delta t + a(N(t + \Delta t) - N(t)) \\ V(0) = \overline{V} \\ if \quad \exists t : V(t) = \theta \to V(t^+) = 0 \end{cases}$$

where $N(t)$ is a Poisson process with activity λ. Suppose that $\lambda \gg \frac{1}{\tau}$ then the model can be simplified to

$$\begin{cases} V(t) = aN(t) \\ V(0) = \overline{V} \\ if \quad \exists t : V(t) = \theta \to V(t^+) = 0 \end{cases}$$

The interspike time of this neuron is T_k, the time of arrival of the k-th spike of the Poisson input, with $k = [\frac{\theta}{a}] + 1$. The density of probability distribution of T_k is a gamma function

$$f_k(t) = \lambda \frac{(\lambda t)^{k-1} e^{-\lambda t}}{(k-1)!} \tag{5.12}$$

which gives the following results

$$\begin{cases} ET_k = \frac{[\frac{\theta}{a}]+1}{\lambda} \\ DT_k = \frac{[\frac{\theta}{a}]+1}{\lambda^2} \\ CV(T_k) \equiv \frac{\sqrt{DT_k}}{ET_k} = \frac{1}{\sqrt{[\frac{\theta}{a}]+1}} \end{cases}$$

Proof.

Without losing generality we can take the initial condition $V(0) = 0$. It is evident that the neuron emits a spike after having received k spikes with $k = [\frac{\theta}{a}] + 1$ since at every spike the potential of the neuron increases an amount a then in order to reach the threshold it needs $[\frac{\theta}{a}]$ spikes where, as usual, $[x]$ means the integer part of x. But it might be that $a[\frac{\theta}{a}] < \theta$ since $[\frac{\theta}{a}]$ is the integer part of $\frac{\theta}{a}$ so that $[\frac{\theta}{a}] + 1$ is the minimum number of spikes arriving at the neuron necessary for the neuron to emit a spike and so the interspike time for this neuron is equal to the time T_k of arrival of the k-th spike. Let us show how to find the density $f_k(t)$. We have to find the $f_k(t)$ such that:

$$\text{Prob}(T_k \in (s+t, s+t+\Delta t)) = f_k(t)\Delta t \qquad (5.13)$$

according to the definition of the density, where s is any initial time, let us choose it to be the instant when the neuron has emitted the last spike. Suppose that $k-1$ spikes arrived in the time interval $(s, s+t)$, the probability of such event being

$$\text{Prob}(N(s, s+t) = k-1)$$

$$= \text{Prob}(N(t+s) - N(s) = k-1) = \frac{(\lambda t)^{k-1} e^{-\lambda t}}{(k-1)!} \qquad (5.14)$$

using the definition of a Poisson process. The probability $\text{Prob}(T_k \in (s+t, s+t+\Delta t))$ can be computed using the independence of the increments of a Poisson process:

$$\text{Prob}(T_k \in (s+t, s+t+\Delta t))$$

$$= \text{Prob}((N(t+s) - N(s) = k-1) \cap (N(s+t+\Delta t) - N(s+t) = 1))$$

$$= \text{Prob}(N(t+s) - N(s) = k-1)\text{Prob}(N(s+t+\Delta t) - N(s+t) = 1)$$

$$= \frac{(\lambda t)^{k-1} e^{-\lambda t}}{(k-1)!} \text{Prob}(N(s+t+\Delta t) - N(s+t) = 1) \qquad (5.15)$$

from the definition of Poisson process we have that:

$$\text{Prob}(N(s+t+\Delta t) - N(s+t) = 1)$$

$$= \frac{(\lambda \Delta t)^1 e^{-\lambda \Delta t}}{1!} \sim \lambda \Delta t(1 - \lambda \Delta t) \sim \lambda \Delta t \qquad (5.16)$$

where we have dropped the terms of the order Δt^2. Substituting the expression found for the term $\text{Prob}(N(s+t+\Delta t) - N(s+t) = 1)$ in equation (5.15) we get the result. The computation of ET_k and DT_k follow in a straightforward way. We remark also that we introduced a new quantity $CV(T_k)$ the *coefficient of variation of* T_k which is often measured in neurobiological experiments. We just give an idea of how to compute ET_k starting from the definition of the density:

$$ET_k = \int_0^\infty t\lambda \frac{(\lambda t)^{k-1} e^{-\lambda t}}{(k-1)!} dt = \frac{k}{\lambda} \int_0^\infty \frac{z^k e^{-z}}{k!} dz \qquad (5.17)$$

where we made the substitution $z = \lambda t$ and the last integral is equal to 1 by well known identities. $\qquad \square$

We generalize the obtained result to the case when the inputs from the external network are two and of different types: one excitatory and one inhibitory. The extension is not trivial at all and we just mention the results and explain, in the next section, how to simulate them. The analytic developments can be found in [Tuckwell (1988)]. Suppose that there are two Poisson inputs to the neuron, $N_1(t)$ with activity λ_1 and $N_2(t)$ with activity λ_2. Assume that $N_1(t)$ is an EPSP and $N_2(T)$ is a IPSP , then we can model the equation of the potential as before assuming that $\lambda_1, \lambda_2 >> \frac{1}{\tau}$

$$V(t+\Delta t) - V(t) = a(N_1(t+\Delta t) - N_1(t)) - b(N_2(t+\Delta t) - N_2(t)) \quad (5.18)$$

with a and b positive. In the small time interval Δt the two processes have the possibility to send at maximum one spike for each of them as we can see from the probability distribution of the Poisson process:

$$\text{Prob}(N(s+t+\Delta t) - N(s+t) - k) - \frac{(\lambda \Delta t)^k e^{-\lambda \Delta t}}{k!} \qquad (5.19)$$

it is evident that the largest contributions to this probability come from the case $k = 0, 1$ where one gets

$$\text{Prob}(N(s+t+\Delta t) - N(s+t) = 0) = \frac{(\lambda \Delta t)^0 e^{-\lambda \Delta t}}{0!} = 1 - \lambda \Delta t \quad (5.20)$$

with $\lambda = \lambda_1, \lambda_2$ and

$$\text{Prob}(N(s+t+\Delta t) - N(s+t) = 1) = \frac{(\lambda \Delta t)^1 e^{-\lambda \Delta t}}{1!} = \lambda \Delta t - (\lambda \Delta t)^2 \sim \lambda \Delta t.$$
$$(5.21)$$

Thus in the small interval of time, supposing $a = b$, the behavior of the potential depends on the values of λ_1, λ_2 if $\lambda_1 > \lambda_2$ the potential depolarizes while if $\lambda_1 < \lambda_2$ the neuron will hyperpolarize. The probability that the potential

$$V(t) = a(N_1(t) - N_2(t))$$

reaches a given threshold θ can be computed in terms of random walk [Tuckwell (1988)] and there might be cases such that the threshold cannot be reached because the inhibitory action is stronger than the excitatory. Let us call T_θ the interspike interval of the examined neuron, the analytic results are:

$$\text{Prob}(T_\theta < \infty) = \left\{ \begin{array}{ll} 1, & \lambda_1 > \lambda_2 \\ (\frac{\lambda_1}{\lambda_2})^\theta, & \lambda_1 < \lambda_2 \end{array} \right\}.$$

These results imply that the interspike interval is finite with probability 1 if the excitatory activity is larger than the inhibitory one and in the opposite case there is a non-zero probability equal to $1 - (\frac{\lambda_1}{\lambda_2})^\theta$ that the interspike interval is infinite, i.e. that there is no spike. These findings are in agreement with the logic: an inhibition larger than the excitation can block the firing of the neurons. The probability density of T_θ is a Bessel function and we do not give it here. In the case $\lambda_1 > \lambda_2$ the expectation and the variance of T_θ are given by the formulas:

$$\left\{ \begin{array}{l} ET_\theta = \frac{\theta}{\lambda_1 - \lambda_2} \\ DT_\theta = \frac{\theta(\lambda_1 + \lambda_2)}{(\lambda_1 - \lambda_2)^3} \\ f = \frac{1}{T_\theta} = \frac{\lambda_1 - \lambda_2}{\theta} \end{array} \right.$$

One can see from these formulas that when $\lambda_1 \to \lambda_2$ the mean value of the interspike time grows in agreement with the previous statements. The quantity f in the last line is the spiking frequency.

5.5 Numeric computation of interspike intervals with Poissonian inputs

The nice analytic results of the previous section cannot be obtained if the hypothesis $\lambda >> 1/\tau$ does not hold because one has to solve the full equation for the change of the potential:

$$V(t + \Delta t) - V(t) = -\frac{V(t)}{\tau}\Delta t + a(N(t + \Delta t) - N(t)) \qquad (5.22)$$

and in this case the simple equation $V(t) = aN(t)$ is not true. In order to find the interspike time one has to solve directly the equation for a given realization of the process $N(t)$ and find the time T which satisfies the threshold condition. By changing the realizations of the process $N(t)$ one gets a sample $T_1, \ldots, T_N$ and from this it is possible, using the methods explained in the previous chapter, to find the mean value and dispersion of the interspike time. The equation for the potential written in the above form can be easily used for this aim:

$$V(t + \Delta t) = V(t) - \frac{V(t)}{\tau}\Delta t + a(N(t + \Delta t) - N(t))$$

knowing the value of the potential at time t one gets immediately the value for the potential at time $t + \Delta t$ once the process $N(t)$ is given. It is simply necessary to start the method at the time $t = 0$ for making the iteration. The possibility of doing this iteration is also connected with the fact that the equation for the potential has been written in the form of the *increments* the variable $V(t)$ and this is particularly useful for the case when the inputs are Poissonian because, as we have seen, also the increments of the process are naturally inserted. This is a particular case of a general form of the differential equations with stochastic terms so it is useful to give a definition of such an equation for understanding better what kind of mathematics we are dealing with.

Definition 5.6.
Given a stochastic process $X(t)$, a stochastic differential equation for the stochastic process $V(t)$ is

$$dV(t) = a(t, X(t), V(t))dt + b(t, X(t), V(t))dX \qquad (5.23)$$

with $dX = X(t + dt) - X(t)$, $dV = V(t + dt) - V(t)$, and a and b are given functions.

Equation (5.22) is a simple case of stochastic differential equation with $a = -\frac{V(t)}{\tau}$, $b = a$, $dt = \Delta t$, $X(t) = N(t)$, with $N(t)$ being a Poisson process. Note that the solution of the equation is also a stochastic process. The iterative procedure is very simple for such a formulation of the differential equations but this way to work on the differential equations has not only been introduced for this reason but also for another important fact: when dealing with stochastic processes the derivative of the processes are not easy to define and to deal with while the increments can be studied more easily. In order to understand how one can simulate a Poisson process $N(t)$ and then realize the numerical estimate of the interspike time we need to recall a result obtained in the previous section. In Lemma 5.2 we have that the time T_k of arrival of the k-th spike after a given time s is distributed according to a probability density given by the gamma distribution defined in (5.12)

$$f_k(t) = \lambda \frac{(\lambda t)^{k-1} e^{-\lambda t}}{(k-1)!} \qquad (5.24)$$

If we put $k = 1$ in this equation with the probability density of arrival of one spike after a given time s, i.e. the probability density of the interval of time between two subsequent spikes:

$$f_1(t) = \lambda \frac{(\lambda t)^{1-1} e^{-\lambda t}}{(1-1)!} = \lambda e^{-\lambda t} \qquad (5.25)$$

Thus the time among the arrival of two spikes is an exponential variable if the process is a Poisson process:

$$\text{Prob}(T_1 \leq t) = \int_0^t \lambda e^{-\lambda t} dt = 1 - e^{-\lambda t} \qquad (5.26)$$

so, according to our discussion of chapter 4, $ET_1 = 1/\lambda$ and $DT_1 = 1/\lambda^2$. This remark allows to construct numerically a Poisson process. If $z_1, \ldots, z_n$ are n variables exponentially distributed with parameter λ then we can construct a Poisson process with the following procedure:

$$
\begin{cases}
N(0) = 0 & 0 \leq t < z_1 \\
N(t) = 1 & z_1 \leq t < z_1 + z_2 \\
\cdots \\
N(t) = n - 1 & z_1 + \cdots + z_{n-1} \leq t < z_1 + \cdots + z_n \\
N(t) = n & t = z_1 + \cdots + z_n
\end{cases}
$$

In this way a function $N(t)$ is defined which has unit jumps $N(t_i) \to N(t_i) + 1$ at the times $t_i = z_1 + \cdots + z_i$ and the intervals of time $t_i - t_{i-1}$ between the jumps are exponentially distributed with parameter λ, so it is

a Poisson process. This procedure together with the iterative solution of
the equation (5.22) is realized in a Matlab program shown in appendix E.
This program has different uses. One is to eliminate the term containing τ
from equation (5.22) and check the results of the previous section, the other
is to obtain the histogram of the interspike time when the decay term has
been inserted. At this point the reader should be enough acquainted with
the Matlab instruction so that he can understand this simple program by
himself without further explanations than the ones given in this section. We
finish this section giving the graphic and numeric output of these Matlab
programs. The first case we are going to show is the simplest of our models,
the one considered in Lemma 5.2, with $V(t) = N(t)$, we set $a = 1$. We show
here the parts of the Matlab program which concerns this simulation and in
Fig. 5.1 the graphic output, the full program can be found in Appendix E.
This simulation checks also the formula for the mean and variance of the
interspike time. The simulation is done in the following way. We construct
a Poisson process with 100 spikes ($number\_spike = 100$), activity $\lambda = 3$
and the spikes of the Poisson process are sent to a neuron with threshold
$\theta = 5$. Then with the loop on i, $i = 1 : 100$ the program constructs 100 interspike times that are put in a vector $t\_interspike\_exc$ of 100 components.
The program then computes the average and variance of this vector and
then verifies if the average and the variance of the interspike times coincide with the formula given in Lemma 5.2, $ET_k = (1 + \theta))/\lambda$ ($a = 1$) and
$CV(T_k) = \frac{1}{\sqrt{\theta+1}}$. The results of the simulation are good because we get
$err = -0.14$ and it must be near to zero from its definition and $err1 = 1.05$
which must be near to 1. The output of Fig. 5.1 is interesting because it
shows all the details of the numerical calculations. We start commenting
the graph from the first picture on the left of the first line. The first graph
shows the exponential random variables constructed in the program, then
its histogram. The third graph are the set of interspike times obtained and
the fourth is their histogram which is of the form of a gamma function in
agreement with Lemma 5.2. The first graph of the second line on the left
is just the graph of the increments of the Poisson process which are taken
to be 1 in this simulation, the second is the graph of the function $N(t)$,
the third is the histogram of the increments of the Poisson process in the
elementary interval Δt of our simulation, and the fourth is the graph of the
gamma distribution of the arrival time of the fifth spike T_5, since we have
increments of 1 and the threshold is 5 it is necessary that 5 spikes of the
Poisson process arrive at the neuron.

```
number_spike=100;
lambda=3;
threshold=5;
for i=1:100

.............

end
m_t_interspike_exc=mean(t_interspike_exc);
var_t_interspike_exc=var(t_interspike_exc);
CV_tinterspike_exc=sqrt(var_tinterspike_exc)/m_tinterspike_exc
err=(1+threshold)/(lambda*m_tinterspike_exc)-1;
err1=sqrt(1+threshold)*CV_t_interspike_exc;
```

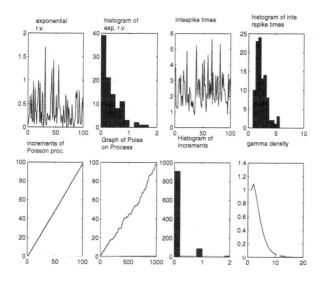

Fig. 5.1   Output of the simulation program for the problem of Lemma 5.2.

The other result of simulations we want to present is the case when there are two input Poisson processes as in equation (5.18). Again the full program is shown in appendix E. In this case the excitatory process has activity $\lambda_1 = 3$ and the inhibitory activity $\lambda_2 = 2$ and the threshold is $\theta = 5$ so the mean of the interspike time $ET_\theta = 5/(3-2) = 5$ and $DT_\theta = 5(3+2)/(3-2) = 25$ according to the formulas shown for this case. In this case the statistics has not been very accurate because the errors on the mean and variance of the interspike interval are rather large, so this program needs to be worked a little bit more but it is a good indication of the direction of the work. We show in Fig. 5.2 the histogram of the interspike interval which is dominated by the exponential decay in agreement with the theory.

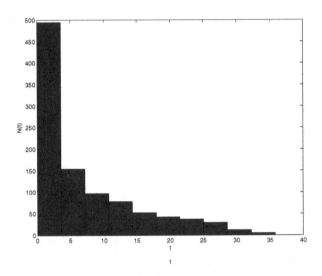

Fig. 5.2   Populations of the interspike times for two input Poisson processes.

We finally show the output of the more general model of a neuron with the decay term and Poissonian input described by equation (5.22) in Fig. 5.3. The sequence of graphs is similar to the sequence of graphs of Fig. 5.1, the only difference is that there is the graph of the time behavior of the potential $V(t)$.

5.6   Neural computation with Brownian inputs

The modelling of inputs of a complex system of neurons to one of its neuron can also be done using other stochastic processes. The use of Poisson processes is natural because it describes very well the discontinuous increments of the membrane potential due to the incoming spikes. In fact we have modelled an incoming spike as a finite amount of charge q arriving at a random time t_i. The inputs were chosen in this way as multiple of this elementary charge. But it is interesting to have the possibility of describing

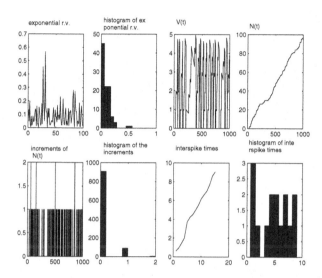

Fig. 5.3 Output of the simulations for the neuron described by the full equation (5.22).

continuous inputs to a neuron and also to have an input signal which can be positive or negative without the necessity of introducing two different processes as we have done in equation (5.18) for modelling the inhibitory and excitatory input. So we describe in this section the neural modelling by means of a process $B(t)$, called *Brownian process* which takes values in $\mathcal{R}$, i.e. continuous values and can get positive or negative values with equal probabilities, so it is useful for cases of synaptic interactions which can have excitatory or inhibitory character with equal probabilities. We are going to give the exact definition of the process $B(t)$ following the same method used for the Poisson process in sections 5.1 and 5.2 so that we can skip the details and the proof of the fact that from the following definition the cylindric probability distributions can be derived.

Definition 5.7.

$B(t)$ is a Brownian process if it satisfies the following conditions:

(1) $B(0) = 0$.
(2) For any $x \in \mathcal{R}$ and any $t, \Delta t \in \mathcal{R}$ we have:

$$\text{Prob}(B(t + \Delta t) - B(t) \in (x, x + dx)) = \frac{e^{-\frac{x^2}{2\Delta t}}}{\sqrt{2\pi}\sqrt{\Delta t}}dx \qquad (5.27)$$

(3) the increments $B(t_{i+1}) - B(t_i)$ corresponding to disjoint intervals $t_{i+1} - t_i$, $i = 1, \ldots, n$ are independent.

Let us give some comments. Originally this process $B(t)$ has been introduced for modelling the position of a particle, belonging to a gas of elastically colliding particles, which starts its motion from the origin of the coordinates and then moves in the medium composed by the other particles of the gas. This explains also the first condition $B(0) = 0$. In other words it was used for modelling the motion of a selected particle in a perfect gas. In this definition we suppose that the gas is one dimensional since we consider the probability of the event $B(t) \in (x, x + dx)$. The second definition gives the probability of such event because we did not give any constraint to the quantities t and Δt appearing in that definition. So from the above definition we get that

$$\mathrm{Prob}(B(t) - B(0) \in (x, x + dx))$$

$$= \mathrm{Prob}(B(t) \in (x, x + dx)) = \frac{e^{-\frac{x^2}{2t}}}{\sqrt{2\pi}\sqrt{t}} dx \qquad (5.28)$$

just putting $t = 0$, $\Delta t = t$. From the properties of the Gaussian distribution that we studied in chapter 4 we have that $EB(t) = 0$ and $EB(t)^2 = t$. This is a well known result that says that the particle oscillates around the origin and so its average position is zero, and its standard deviation from it is proportional to the elapsed time t. So the particle will have equal probability to stay at the left or at the right of the origin, i.e. to take positive and negative values. This is a consequence of the Gaussian density of the definition. These properties can be expressed in the language of input signals to the neuron. The value of the input signal will oscillate around a mean value, zero in this case, but its mean square value will be proportional to the time t. Note also that it is implicit from the definition that the signal gets continuous values since x is a real variable. The equation of the evolution for the $I\&F$ model with Brownian inputs take the simple form described in the following definition:

Definition 5.8.

The $I\&F$ model with Brownian inputs $B(t)$ is defined by the

$$\begin{cases} dV(t) = -\frac{V(t)}{\tau} dt + \sigma dB(t) \\ V(0) = \overline{V} \\ \text{if} \quad \exists t : V(t) = \theta \rightarrow V(t^+) = 0 \end{cases}$$

where $dB(t) = B(t + dt) - B(t)$ are the increments of the Brownian process $B(t)$ and σ is a scaling constant.

Another characteristic of the Brownian process is that there are many interesting analytic results about the probability density of the interspike time t of this definition. Since this is an elementary introduction we do not have the space and possibility for giving the proof of these interesting results, so we simply give the results and their meaning without proofs. The proofs can be found in [Tuckwell (1988)] and in the rich literature given there. Moreover, there are many important papers on probability theory where the questions connected with the problems of finding the density of the interspike time are solved in practice. They just use the terminology of the first exit time or passage time through a threshold. From the definition of the Brownian process it is easy to get the DF of the $B(t)$:

$$\text{Prob}(B(t) < x) = \int_{-\infty}^{x} \frac{e^{-\frac{u^2}{2t}}}{\sqrt{2\pi}\sqrt{t}} du \qquad (5.29)$$

and all the discussion of the properties of this distribution has been done already in section 4.5 and so we will not repeat it here. We just give the result on the interspike time T_θ for a model simpler than the *I&F* model:

$$\begin{cases} dV(t) = \mu dt + \sigma dB(t) \\ V(0) = \overline{V} \\ \text{if } \exists T_\theta : V(T_\theta) = \theta \to V(T_\theta^+) = 0. \end{cases}$$

Solving a nice analytic equation for the interspike probability density density (the *renewal equation*) which can be derived for stochastic differential equations with Brownian motion, the probability density $f_\theta(t)$ for T_θ can be obtained

$$f_\theta(t) = \frac{\theta - \overline{V}}{\sqrt{2\pi\sigma^2 t^3}} e^{-\frac{(\theta - \overline{V} - \mu t)^2}{2\sigma^2 t}} \qquad (5.30)$$

with $t > 0$ and $\theta > \overline{V}$. From this density it is possible to derive the moments of the interspike time T_θ:

$$ET_\theta = \frac{\theta - \overline{V}}{\mu} \qquad (5.31)$$

$$DT_\theta = \frac{(\theta - \overline{V})\sigma^2}{\mu^3} \qquad (5.32)$$

These expressions make sense only in the case $\theta > \overline{V}$ as it should be.

PART 2
Clustering

Chapter 6

An introduction to clustering techniques and self-organizing algorithms

6.1 A brief overview of clustering technique

The clustering problem has been addressed in many contexts and by researchers in many disciplines; this reflects its broad appeal and usefulness as one of the steps in exploratory data analysis. Clustering is the process of grouping together similar entities. It can be applied to any data: genes, samples, time points of a time series, etc. The particular type of input makes no difference to the clustering algorithm. The algorithm will treat all inputs as a set of n numbers or n-dimensional vectors. The definition of group, *cluster*, is not precisely defined. In many different applications, the best definition depends on the type of data and the desired results. Several working definitions of clusters are commonly used:

Definition of well-separated clustering: a cluster is a set of points such that any point in a cluster is closer (or more similar) to every other point in the cluster than to any point not in the cluster.

Definition of center-based cluster: a cluster is a set of objects such that an object in a cluster is closer to the *center* (centroid) of the cluster, than to the center of any other cluster.

Definition of contiguous cluster: a cluster is a set of points such that a point in a cluster is closer to one more other points in the cluster than to any point not in the cluster.

Definition of density-based cluster: a cluster is a dense region of points, which is separated by low-density regions, from other regions of high density.

Definition of similarity-based cluster: a cluster is a set of objects that are *similar* and objects in other clusters are not *similar*.

If one wants to group together things that are similar, one should start by defining the meaning of similarity. In other words, we need a very precise measure of similarity. Such a measure of similarity can be considered as a distance or metric. A distance is a formula that takes two points in the input space of the problem and calculates a positive number that contains information about how close the two points are to each other. There are many different ways in which such a measure of similarity can be calculated. The final result of the clustering depends strongly on the choice of the formula used.

6.2   Distance metric

Definition 6.1. A distance metric d is a function that takes as argument two points x and y in an n-dimensional space $\mathbb{R}^n$ and has the following properties:

(1) $d(x, y) \geq 0$, non-negativity.
(2) $d(x, y) = 0$ if and only if $x = y$, distance of an object to itself is 0.
(3) $d(x, y) = d(y, x)$, symmetry.
(4) $d(x, y) \leq d(x, z) + d(z, y)$, where $z \in \mathbb{R}^n$; triangle inequality.

Anderberg [Anderberg (1973)] gives a thorough review of measures and metrics, also discusses their interrelationships. Note that any metric is also a measure, while a measure is not always a metric.

Here, we give a brief description of measures:

- Minkowski Distance, defined as:

$$d(x, y) = (\sum_{i=1}^{N} |x_i - y_i|^q)^{1/q} \tag{6.1}$$

  where q is a positive integer.
- Euclidean Distance, defined as:

$$d(x, y) = \sqrt{\sum_{i=1}^{N} (x_i - y_i)^2} \tag{6.2}$$

  Note that this is equal to the Minkowski distance for $q = 2$.
- Manhattan Distance, defined as:

$$d(x, y) = \sum_{i=1}^{N} |x_i - y_i| \tag{6.3}$$

  Note that this is equal to the Minkowski distance for $q = 1$.

- Maximum Distance, defined as:

$$d(x, y) = max_{i=1}^{N}|x_i - y_i| \tag{6.4}$$

Note that this is equal to the Minkowski distance for $q \to \infty$.

- Mahalanobis distance, defined as:

$$d(x, y) = \sqrt{(x - y)^T S^{-1}(x - y)} \tag{6.5}$$

where S is any $n \times n$ positive definite matrix and $(x - y)^T$ is the transposition of $(x - y)$. The role of the matrix S is to distort the space as desired. S is usually the covariance matrix of the data set.

- Angle distance, defined as:

$$d(x, y) = \cos(\theta) = \frac{x \cdot y}{\| x \| \| y \|} \tag{6.6}$$

where $x \cdot y$ [1] is the dot product and $\| \cdot \|$ [2] is the norm.

- Correlation distance, defined as:

$$d(x, y) = 1 - r_{xy} \tag{6.7}$$

where r_{xy} is the Pearson correlation coefficient of the vectors x and y:

$$r_{xy} = \frac{s_{xy}}{\sqrt{s_x}\sqrt{s_y}} = \frac{\sum_{i=1}^{N}(x_i - \bar{x})(y_i - \bar{y})}{\sqrt{\sum_{i=1}^{N}(x_i - \bar{x})^2}\sqrt{\sum_{i=1}^{N}(y_i - \bar{y})^2}} \tag{6.8}$$

Note that since $-1 \leq r_{xy} \leq 1$, the distance $1 - r_{xy}$ will take values between 0 and 2.

- The squared Euclidean distance, defined as:

$$d(x, y) = \sum_{i=1}^{N}(x_i - y_i)^2 \tag{6.9}$$

A comparison of distances

(1) Euclidean distance, a special case of the Minkowski metric, has an intuitive meaning as it is commonly used to evaluate the proximity of objects in two or three-dimensional space. It works well when a data set has *compact* or *isolated* clusters.

(2) Squared Euclidean distance tends to emphasize the distances. The same data clustered with squared distance might appear less compact.

---

[1] $x \cdot y = \sum_{i=1}^{N} x_i y_i$

[2] $\| x \| = \sqrt{\sum_{i=1}^{N} x_i^2}$

(3) Angle distance takes into account only the angle, not the magnitude.
(4) Correlation distance has the advantage of calculating similarity depending only on the pattern but not on the absolute magnitude of the spatial vector.
(5) Manhattan distance does not match the similar set constructed with the Euclidean distance.
(6) Mahalanobis distance can warp the space in any convenient way.

6.3 Clustering algorithms

There are many clustering methods available, and each of them may give a different grouping of a data set. The choice of a particular method will depend on the type of output desired. In general, clustering methods may be divided into two categories based on the cluster structure which they produce:

- Hierarchical methods
- Non-hierarchical methods

6.4 Hierarchical methods

The hierarchical methods produce a set of nested clusters in which each pair of objects or clusters is progressively nested in a larger cluster until only one cluster remains. The hierarchical methods can be further divided into agglomerative or divisive methods. In agglomerative methods, the hierarchy is built up in a series of $N - 1$ agglomerations of pairs of objects, beginning with the un-clustered data set. The result of a hierarchical clustering algorithm can be graphically displayed as a tree, called dendogram. This tree graphically displays the merging process and the intermediate clusters. Thereby it can be used to divide the data into pre-determined number of clusters. The division may be done by cutting the tree at a certain depth (distance from the root). The dendogram in Fig. 6.1 shows how points can be merged into a single cluster.

The procedure of agglomerative algorithms is the following:

(1) Place each element of the given set S in its own cluster (singleton), creating the list of clusters L:

$$L = S_1, S_2, S_3, ..., S_{n-1}, S_n$$

(2) Compute a merging cost function between every pair of elements in L to find the two closest clusters (S_i, S_j) which will be the cheapest couple to merge.
(3) Remove S_i and S_j from L.
(4) Merge S_i and S_j to create a new internal node S_{ij} which will be the parent of S_i and S_j in the result tree.
(5) Repeat from step 2 until only a single cluster remains.

The major difference among the agglomerative algorithms is the definition of cost of merging two sets S_i, S_j (see Table 6.1).

Table 6.1   *Hierarchical clustering* models

Method	Cost function
Single link	$\min_{x_i \in S_i, x_j \in S_j} d(x_i, x_j)$
Average link	$\frac{1}{|S_i||S_j|} \sum_{x_i \in S_i} \sum_{x_j \in S_j} d(x_i, x_j)$
Complete link	$\max_{x_i \in S_i, x_j \in S_j} d(x_i, x_j)$

The less common divisive methods begin with all objects in a single cluster and at each of $N - 1$ steps divides some clusters into two smaller clusters, until each object resides in its own cluster.

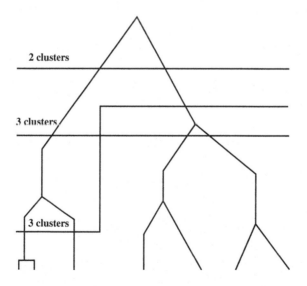

Fig. 6.1   Dendogram.

A few conclusions can be drawn: hierarchical clustering suffers from the fact that the data objects are grouped based initially on clusters chosen almost randomly, with no guarantee of global optimization. This problem is even exacerbated by the deterministic nature of hierarchical clustering which implies a critical dependence on the initial choices of the algorithm. Hierarchical clustering has also been noted by statisticians to have trouble with one or more of the following: lack of robustness, non-uniqueness, non-convex shape, and a tendency to break large clusters.

6.5 Non-hierarchical methods

The non-hierarchical methods divide a data set of N objects into M clusters, with or without overlap. These methods are sometimes divided into partitioning methods, in which the classes are mutually exclusive, and the less common clumping method, in which overlap is allowed. Each object is a member of the cluster with which it is most similar, however the threshold of similarity has to be defined.

In contrast to hierarchical techniques, these clustering techniques create a one-level partitioning of the data points. If K is the desired number of clusters, then partitional approaches typically find all K clusters at once.

There are a number of partitional techniques, but we shall only describe the K-means algorithm which is widely used in document clustering. K-means is based on the idea that a center point can represent a cluster. In particular, for K-means we use the notion of centroid, which is the mean or median point of a group of points.

The K-mean algorithm is derived by asking how we can obtain a partition of the data which optimizes the squared error criterion, which works well enough with isolated and compact clusters. The squared error for a cluster S_i is:

$$c(S_i) = \sum_{r=1}^{|S_i|} d(\overline{x}^i, x_r^i) \tag{6.10}$$

where $\overline{x}^i$ is the centroid.
The procedure of K-means is:

(1) Select randomly K points as the initial centroids.
(2) Assign all points to the closest centroid.
(3) Evaluate the new centroid of each cluster.

(4) Repeat steps 2 and 3 until the centroids do not change or the squared error ceases to decrease significantly.

A major problem with this algorithm is that it is sensitive to the selection of the initial partition and may converge to a local minimum of the cost function if the initial partition is not properly chosen. Several variants [Ben-Dor (1999)] of the K-means algorithm have been reported in the literature. Some of them attempt to select a good initial partition so that the algorithm is more likely to find the global minimum value. Another variation is to permit splitting and merging of the resulting clusters. Typically, a cluster is split when its variance is above a pre-specified threshold, and two clusters are merged when the distance between their centroids is below another pre-specified threshold. Using this variant, it is possible to obtain the optimal partition starting from any arbitrary initial partition, provided proper threshold values are specified.

6.6 Graph-theoretic clustering

We will introduce some terms and results of graph theory, which we will use in this section. For more information on concepts and standard algorithms of graph theory, we propose the book [Gondran (1979)].

- A graph G is a couple (V, E), where $V = 1, ..., n$ is the vertex set, $E \subseteq V \times V$ is the edge set.
- A weight function of G is an application $w : E \rightarrow \mathbb{R}_+$ which associate an edge $e \in E$ with a real positive number $w(e)$.
- A loop is an edge whose end vertices are the same vertex.
- A graph is acyclic if it contains no cycles.
- An edge is multiple if there is another edge with the same end vertices; otherwise it is simple.
- A subgraph of a graph G is a graph whose vertex and edge sets are subsets of those of G.
- A tree is a connected acyclic simple graph.
- A subgraph H is a spanning subgraph of a graph G if it has the same vertex set as G.
- A spanning tree is a spanning subgraph that is a tree.
- A complete graph is a graph in which each pair of graph vertices is connected by an edge. The complete graph with n graph vertices is denoted K_n.

- A clique of a graph is its maximal complete subgraph, although some authors define a clique as any complete subgraph.
- Two subsets A and B of V, such that $A \cup B = V$ and $A \cap B = \emptyset$, define a cut in G, which we denote as (A, B). The sum of the weights of the edges crossing the cut defines its value. We use a real function c to represent the value of a cut. So, cut (A, B) has value $c(A, B)$. In fact, function c can also be applied when A and B do not cover V, but $A \cup B \subset V$.
- For graph G, there exists a weighted graph T_G, which we call the minimum cut tree of G. The min-cut tree is defined over V and has the property that we can find the minimum cut between two nodes s, t in G by inspecting the path that connects s and t in T_G.

In some clustering graphs the data to be clustered are represented as an undirected edge-weighted graph with no self-loops $G = (V, E, w)$, w is the weight function. Vertices in G correspond to data points, edges represent neighborhood relationships, and edge-weights reflect similarity between pairs of linked vertices. The graph G is represented by the corresponding weighted adjacency (or similarity) matrix, which is the $n \times n$ symmetric matrix $A = (a_{ij})$ defined as:

$$a_{i,j} = \left\{ \begin{array}{cc} w(i,j), & \text{if } (i,j) \in E \\ 0, & \text{otherwise} \end{array} \right\}$$

We do not assume that the similarity weights obey the triangular inequality, and self similarities $w(i, i)$ are not defined.

For other clustering graphs, the data are represented as an un-weighted graph G' built fixed a threshold s and assuming that two edges are connected if and only if their similarity is above the threshold:

$$(i, j) \in E(G') \Leftrightarrow s_{i,j} > s$$

There are two main graph-theoretic clustering methods:

- the one based on construction of the minimal spanning tree (MST) (built on minimum cut) of the data, and then deleting the MST edges with the largest lengths to generate clusters. A minimum spanning tree is a spanning tree with weight less than or equal to the weight of every other spanning tree.
- the one based on construction of the maximal clique. A clique C of G is maximal if no external vertex to C is connected with more than $|C| - 1$ vertices of C, where $|C|$ is the number of vertices in C.

There are many techniques based on graph approach but we shall only describe the CAST algorithm which is widely used in document clustering.

6.7 CAST

CAST was elaborated by Ben-Dor et al. [Ben-Dor (1999)] introducing the idea of a corrupted clique graph data model.

The input data is assumed to come from the underlying cluster structure by *contamination* with random errors caused by the complex process of measurements. To be specific, it is assumed that the true clustering of the data points can be represented by a clique graph H. The similarity graph G is derived from H by flipping each edge/non-edge with probability a. Therefore, to cluster a data set is the same as identifying the original clique graph H given the corrupted version G with as few flips (errors) as possible.

The input to the algorithm includes the pairwise similarities of the genes, and a cutoff parameter t (which is a real number between 0 and 1). The clusters are constructed one at a time. The current cluster under construction is called C_{open}. The affinity of a datum g, $a(g)$, is defined to be the sum of similarity values between g and all the genes in C_{open}. A datum g is said to have high affinity if $a(g) \geq t|C_{open}|$. Otherwise, g is said to have low affinity. Note that the affinity of a datum depends on the data that are already in C_{open}. The algorithm alternates between adding high affinity genes to C_{open}, and removing low affinity data from C_{open}. C_{open} is closed when no more data can be added to or removed from it. Once a cluster is closed, it is not considered any more by the algorithm. The algorithm iterates until all the data have been assigned to clusters and the current C_{open} is open.

A few considerations: the obvious advantage of CAST algorithm is that the number of clusters is determined by the algorithm and no prior knowledge of the cluster structure is required. However it is also important to note that the affinity threshold and error rate indirectly influence the cluster structure: increasing the threshold clusters will be smaller until they are reduced at one element for a sufficient high threshold. On the other hand, if the threshold decreases the clusters will be larger until they are a partition with one cluster. Therefore the choice of the threshold is equivalent to the choice of the level cut in the dendograms of agglomerative algorithms.

It is important to note that the algorithm can be controlled at each step; but there is no normal proof for the convergence of the algorithm.

We have introduced briefly several different clustering algorithms, all of which partition the data set from different points of views and tend to emphasize different types of regularities in the data. As a consequence, different clustering algorithms or even the same algorithm along with different input parameters may yield very different partitions given the same data set. Because of these differences having a good control on the properties of clustering algorithms is very important.

In this chapter we have focused only on Kohonen algorithm (or Kohonen neural network) [Kohonen (1989)]-[Kohonen (1991)] since its properties are well known and can be controlled by the user.

6.8   The Kohonen network

An intuitive description

The Kohonen network [Kohonen (1989)]-[Kohonen (1991)] is formed by a single layered neural network. The data are presented to the input and the output neurons are organized with a simple neighborhood structure. Each neuron is associated with a reference vector (the weight vector), and each data point is *mapped* to the neuron with the *closest* (in the sense of the Euclidean distance) reference vector. In the process of running the algorithm, each data point acts as a training sample which directs the movement of the reference vectors towards the value of the data of this sample. The vectors associated with neurons, called weights, change during the learning process and tend to the characteristic values of the distribution of the input data. The characteristic value of one cluster can be intuitively understood as the typical value of the data in the cluster and will be defined more precisely in the next subsections. At the end of the process the set of input data is partitioned in disjoint sets (the clusters) and the weight associated with each neuron is the characteristic value of the cluster associated with the neuron in one dimensional case, which is the case of interest to us. We limit our analysis to this case because the condition of convergence of the algorithm is easier to check, the clusters of the partition are easier to visualize and it is not difficult to compare the behavior of the genes in the clusters corresponding to the different biological conditions. Each neuron individuates one cluster, the physical or biological entities with measure values belonging to the same cluster are considered to be involved in the same cellular process. Thus the genes with expressions belonging to the same cluster might be functionally related.

The following points show the main properties which make the Kohonen network useful for clustering :

(1) Low dimension of the network and its simple structure.
(2) Simple representation of clusters by means of vectors associated with each neuron.
(3) Topology of the input data set is somehow mapped in the topology of the weights of the network.
(4) Learning is unsupervised.
(5) Self-organized property.

Points 1)-2) are simple to understood and many examples are shown afterward. Point 3) means that neighboring neurons have weight vectors not very different from each other. Point 4) means that there is no need to have an external constraint to drive the weights towards their right values beside the input to the network and that the learning process finds by itself the right topology and the right values. This holds only if the learning process with which the network is constructed converges almost everywhere (a.e.), i.e. independently from initial conditions and from order of data insertion, or if the mean values are taken. The self-organization is formulated in the current literature referring to some universality of the structure of the network for a given data set. It is connected to point 3) and is also a consequence of a.e. convergence or of the convergence of the weights to the same limit over many different learning processes.

Exact definition

In this subsection we give the definitions using exact mathematical terms. We restrict ourselves to the particularly simple one-dimensional case which is the most interesting for our applications. First we show how the Kohonen network is used for classification and then what is the process of its construction.

Let $I = I_1, ..., I_N$ be a partition of the interval $(0, A)$ of the possible values of the expression levels in the intervals I_i, $\bigcup_i I_i = (0, A)$ and $I_i \bigcap I_j = 0$. Suppose that the construction of the Kohonen network has already been done and the I_i are the clusters. Let ω_i, $i = 1, ..., N$ be the weights or the characteristic values of the clusters which will be exactly defined below. Then a datum ξ is said to have the property i if $\xi \in I_i$. After giving these concepts we can define the classification error.

Definition 6.2. The classification error is

$$|\xi - \omega_i|$$

Then the global classification error E of the network is

$$E = \frac{1}{T} \sum_{i=1}^{N} \int_{I_i} \|\xi - \omega_i\|^2 f(\xi)\, d\xi \qquad (6.11)$$

where $f(\xi)$ is the density of probability distribution of the input data and $T = \sum_{i=1}^{N} |I_i|$, with $|I_i|$ is the number of data in the set I_i.

The partition $I = I_1, ..., I_N$ is optimal if the associated classification error E is minimal. The characteristic vectors ω_i are the values which minimize E. Before giving exact definitions let us explain in simple terms the procedure for determining the sets I_i and the associated weights ω_i. Let $x(1), ..., x(P)$ be a sequence of values randomly extracted from the data set, distributed with the density $f(x)$ and take randomly the initial values $\omega_1(0), ..., \omega_N(0)$ of the weights. When an input pattern $x(n)$, $n = 1, ..., P$, is presented to the network all the differences $|x(n) - \omega_i(n-1)|$ are computed and the winner neuron is the neuron j with minimal difference $|\xi(n) - \omega_j(n-1)|$. The weight of this neuron is changed in a way defined below, or, in some cases, the weights of the neighboring neurons are changed. Then this procedure is repeated with another input pattern $x(n+1)$ and with the new weights $\omega_i(n)$ until the weights $\omega_i(n)$ converge to some fixed values for P large enough.

In this way we get a random sequence $\omega(n) = (\omega_1(n), ..., \omega_N(n))$ which converges a.e., under suitable conditions on the data set, with respect to the order of input data and the random choice of the initial conditions of the weights, n is the number of iterations of the procedure. The learning process is the sequence $\omega(n)$ and the S.O. (self-organizing property) coincides in practice with the almost everywhere convergence of $\omega(n)$. The learning process converges somehow to the optimal partition in the Kohonen algorithm. In fact the algorithm can, with some approximation, be viewed as a gradient method applied to the function E:

$$\omega_i(n+1) = \omega_i(n) + \eta(n)[\xi(n) - \omega_i(n)]$$
$$\sim \omega_i(n) - \frac{1}{2}\eta(n)\nabla_{\omega_i(n)} E$$

In one dimension the Kohonen algorithm in the simplest version of the winner-take-all case is :

(1) Fix N.
(2) Choose randomly at the initial step ($n = 0$) the ω_i ($0 \leq i \leq N$).
(3) Extract randomly the data $\xi(1)$ from the data set.
(4) Compute the modules

$$|\omega_i(0) - \xi(1)| \quad i = 1, \ldots, N$$

(5) Choose the neuron v such that

$$|\omega_v(0) - \xi(1)|$$

is the minimum distance. v is the winner neuron.
(6) Update only the weight of the winner neuron:

$$\omega_v(1) = \omega_v(0) + \eta(1)(\xi(1) - \omega_v(0)).$$

(7) $n = n + 1$.

One of the basic properties of the Kohonen network is that the weights are ordered if the learning process converges.

Definition 6.3.

A one-dimensional configuration $\omega_1, \ldots, \omega_N$ is ordered if:

$$|r - s| < |r - q| \Leftrightarrow |\omega_r - \omega_s| < |\omega_r - \omega_q|, \ \forall r, s, q \in \{1, 2, \ldots, N\}$$

the order holds also for the other inequality $|\omega_r - \omega_s| > |\omega_r - \omega_q|$.

Then we have the following lemma:

Lemma 6.1. *The ordering property.*

If the Kohonen learning algorithm applied to one-dimensional configuration of weights converges, the configuration orders itself at a certain step of the process. The same order is preserved at each subsequent step of the algorithm.

This property allows to check when the algorithm converges: if the final configuration of weights is not ordered at the end of the learning process, then the algorithm does not converge. We also check the remarkable property proved by Kohonen ([Taylor (1995)], [Tolat (1990)], [Vesanto (2000)]) that the mean process $\omega(n)$, i.e. the process obtained by averaging with respect to different choices of the sequence $x(1), \ldots, x(P)$ is always converging. But to obtain the a.e. convergence from the convergence of the mean values, additional hypothesis must be used and the discussion and the applications of these results to a case of genetic classification is the main topic of this section.

General formulation

We describe now the Kohonen algorithm in more general terms for allowing the treatment of all the possible cases.

The Kohonen network is composed of a single layer of output units $\mathcal{O}_i$, $i = 1, ..., N$ each being fully connected to a set of inputs $\xi_j(n)$, $j = 1, ..., M$. An M-dimensional weight vector $\omega_i(n)$, $\omega_i(n) = (\omega_{ij}(n), j = 1, ..., M)$ is associated with each neuron. n indicates the n-th step of the algorithm.

We assume that the inputs $\xi_j(n)$, $j = 1, ..., M$ are independently chosen according to a probability distribution $f(x)$. For each input $\xi_j(n)$, $j = 1, ..., M$ we choose one of the output units, called the *winner*. The *winner* is the output unit with the smallest distance between its weight vector $\omega_v(n)$ and the input

$$||\omega_v(n-1) - \xi(n)||$$

where $||.||$ represents Euclidean norm. Let $\overline{I}(.,.)$ be the function

$$\overline{I}(\omega_v(n), \xi(n+1)) = I_{\{||\omega_v(n)-\xi(n+1)|| < ||\omega_j(n)-\xi(n+1)||, \ j \neq v\}}$$

where I_A is the characteristic function of the event A, i.e. $I_A(x) = 1$ if $x \in A$ and $I_A(x) = 0$ if $x \notin A$.

This function selects the event in which the weight of the neuron v is nearest to the input data $\xi(n)$ and it is necessary for writing the learning process in a compact form. The generalized Kohonen algorithm updates the weights of the neurons belonging to a given neighbor of the *winner* neuron:

$$\omega_{ij}(i+1) = \omega_{ij}(n) + \eta(n)\Gamma(i,v)\overline{I}(\omega_v(n), \xi(n+1)) \ (\xi_j(n+1) - \omega_{ij}(n)) \quad (6.12)$$

$i = 1, ..., N$ and $j = 1, ..., M$ or in vector form

$$\omega_i(n+1) = \omega_i(n) + \eta(n)\Gamma(i,v)\overline{I}(\omega_v(n), \xi(n+1)) \ (\xi(n+1) - \omega_i(n)) \quad (6.13)$$

where $\eta(n)$ is the positive learning parameter $\eta(0) < 1$, $\eta(n) \geq \eta(n+1)$ and $\Gamma(i,v)$ is a non-increasing function of $|i - v|$, the distance among the neuron i and v on the lattice where the neurons of the network are located.

This version is more general than the *winner-take-all* rule explained before. Not only the weight of the winner neuron is updated but also the weights of the neurons which belong to a neighborhood defined by the function $\Gamma(i,v)$. We discuss various choices of the function $\Gamma(i,v)$ below. After the learning procedure is finished, the set of input vector will be partitioned into non-overlapping clusters. This means that a new signal $\xi(n+1)$ is classified as the pattern i if and only if

$$||\omega_i - \xi(n+1)|| \leq ||\omega_j - \xi(n+1)||, \ j \neq i.$$

Let us introduce the definition of Voronoi tessellation $\Pi(y) = (\Pi(y)_i, i = 1, ..., N)$ associated with a family vectors $y_1, ..., y_N \in \Omega$, Ω being a given compact of $\mathbb{R}^M$.

Definition 6.4.

For a given compact subset $\Omega \in \mathbb{R}^M$, the Voronoi tessellation $\Pi(y) = (\Pi(y)_i, \quad i = 1, ..., N)$ associated with a family of vectors $y_1, ..., y_N$ is the partition of Ω:

$$\Pi(y)_i = \{x, \|y_i - x\| \leq \|y_j - x\|, j \neq i\} \quad i = 1, ..., N. \tag{6.14}$$

Therefore a Voronoi cell of a unit i contains those vectors which are closer to the weight ω_i than to the other weights. The characteristic values mentioned before are the limit of the sequences of the vectors $\omega_i(n)$ defined by the above algorithm and are weights of the Voronoi tessellation obtained in the limit.

A crucial point of the algorithm is the choice of the neighborhood function $\Gamma(i, v)$ of the *winner* neuron. It determines the region around the *winner* neuron where there are the neurons which update their weight vectors together with the winner neuron. A convenient choice is the finite region of activation of the winner neuron, i.e. $\Gamma = \Lambda$ where:

$$\Lambda(i, v) = \begin{cases} 1 \text{ if } |i - v| \leq s \\ 0 \text{ otherwise} \end{cases} \tag{6.15}$$

where $|.|$ represents the distance between the neuron i and the *winner* neuron v.

If $s = 1$ and the neural network is one dimensional, the region of activation includes the *winner* and the two nearest units (see Fig. 6.2); if the network is designed in two dimensions then the range includes the eight nearest neighbor units near the *winner*.

If $\Lambda(i, v) = \delta_{iv}$, the algorithm will coincide with the *winner-takes-all* algorithm we described in the previous section.

Another choice, often used in the applied research, for the neighborhood function Γ is a Gaussian function h defining a region around the winner neuron with amplitude decreasing with the number of iterations of the learning process:

$$h(i, v, n) = \exp\left(-\frac{|i - v|^2}{\sigma(n)^2}\right) \tag{6.16}$$

where $\sigma(n)$ is a decreasing function. A commonly used choice is:

$$\sigma(n) = \sigma_i \left(\frac{\sigma_f}{\sigma_i}\right)^{\frac{n}{n_{max}}}$$

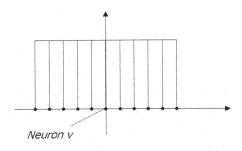

Fig. 6.2   Neighborhood function $\Lambda(i, v)$.

where n_{max} is the maximum number of iterations of the algorithm and σ_f, σ_i are respectively the final and initial value of the parameter σ (see Fig. 6.3).

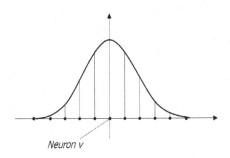

Fig. 6.3   Neighborhood function $h(i, v, n)$.

6.9   Numerical investigations and applications of Kohonen algorithm

Numerical studies

In this section we illustrate our numerical simulations about the convergence of the Kohonen algorithm. First we consider a uniformly distributed data set, then a normally distributed data set, all the data are one-dimensional as we have already mentioned.

We see that the algorithm does not even converge in mean (and so also not a.e.) if:

(1) $\eta(n)$, the learning parameter, decreases too fast.
(2) The neighborhood functions ($\Lambda(i, v)$ (6.15), or $h(i, v, n)$ (6.16)) have a range of action too small or too large.

In addition, although the learning parameter and the neighborhood function are optimally chosen, the convergence of the algorithm is slow and it needs a large number of iterations in order to have a good accuracy. So, when the data set is not large enough, it is useful to repeat the presentation of data several times in random order until we have a large data set.

In particular in the case of uniformly distributed data, chosen inside the interval $[0, 1]$, we verify numerically that, having a large data set, choosing any neighborhood function and using as learning parameter $\eta(n) = \frac{1}{n^\alpha}$ with $\alpha \geq 1$ the algorithm does not converge in any sense (for different initial choices of weights we have different outputs) and the weights are not ordered during the learning procedure. Instead using $\eta(n) = \frac{1}{n^\alpha}$, with $\alpha \leq \frac{1}{3}$, we have the convergence in mean. So the convergence property depends on the velocity of decay of $\eta(n)$. In fact if $\eta(n)$ decreases too fast, e.g. $\eta(n) = \frac{1}{n^\alpha}$, with $\alpha \geq 1$, the updated weights change their values very little during the learning and so the algorithm is not able to find the final configuration of weights. $\eta(n) \sim 1/n$ is too fast a decay because after 100 iterations the variation of the weights is already very small and so there is no convergence while $\eta(n) \sim \frac{1}{\sqrt[3]{n}}$ decreases less quickly (it assumes values less than 0.01 from $n > 10^6$) and its velocity of decrease is sufficient to have the convergence.

The choice of $\eta(n)$ is basic not only for the convergence but also for accuracy. In fact we can have the convergence of the algorithm though the algorithm is not able to identify all the limit weights but only some of them. This happens when the weights are updated too fast in the last part of the learning procedure or when the range of $\eta(n)$ does not cover all the interval $(0, 1)$, for example when the range of $\eta(n)$ is $(0.5, 1)$. We analyzed the following $\eta(n)$:

(1) $\eta(n) = \frac{1}{\sqrt{log(n)}}$
(2) $\eta(n) = \frac{1}{log(n)}$
(3) $\eta(n) = \eta_i \left(\frac{\eta_f}{\eta_i}\right)^{\frac{n}{n_{max}}}$
(4) $\eta(n) = \frac{1}{\sqrt[3]{n}}$

(where η_i and η_f are respectively the initial and final values of the function η and n_{max} the maximum number of iterations). For all these cases we have convergence in mean, but for each case there is a different accuracy.

Choosing

(1) $\eta(n) = \eta_i(1 - \frac{n}{n_{max}})$

(2) $\eta(n) = \frac{\sqrt{6*log(n)}}{\sqrt{(n)}+1}$

we have convergence a.e. The values of the constants and the particular forms of the functions $\eta(n)$ have been determined for satisfying the constraint $0 \leq \eta(n) \leq 1$.

Before explaining the reasons of this statement, we want to discuss the connection of the convergence with the values of the parameters. The choices of the parameters depend on the data distribution. For example for case 3), in the case of uniformly and normally distributed data, generally we have convergence if we choose η_i between 0.1 and 0.9 and η_f between 10^{-6} and 0.1; in case 1) of the second list the range of η_i is $(0.1, 0.9)$. Instead for example with log-normal distributed data the range of η_i is $(0.4, 0.9)$ and of η_f is $(0.01, 0.1)$ in case 3) and in the other case the range of η_i is $(0.4, 0.8)$.

After many simulations we saw that there is convergence in mean for $\eta(n)$ such that:

$$\frac{1}{\sqrt[3]{n}} \leq \eta(n) \leq \frac{\sqrt{6 * log(n)}}{\sqrt{(n)} + 1},$$

while there is a.e. convergence for $\eta(n)$ such that

$$\frac{\sqrt{6 * log(n)}}{\sqrt{(n)} + 1} \leq \eta(n) \leq \eta_i(1 - \frac{n}{n_{max}})$$

The convergence depends also on the values of parameters concerning the neighborhood function, that is the range of the action of the winner neuron which is determined by s in the case of $\Lambda(i, v)$, and by σ_i and σ_f in the case of $h(i, v, n)$. The choice of s depends strongly on the number of weights we fix at the beginning and the number of iterations. For example using a data set of about 10000 uniformly distributed data if we choose $\eta(n) = \frac{\sqrt{6*log(n)}}{\sqrt{(n)}+1}$, $\Lambda(i, v)$ with $s = 1$ as neighborhood function and we want to find 30 groups we do not have the convergence (the weights are not ordered) but changing the value of s conveniently (in this case $s \geq 2$) we

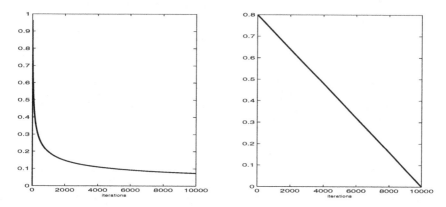

Fig. 6.4   On the left we have $\eta(n) = \frac{\sqrt{6*log(n)}}{\sqrt{(n)}+1}$, on the right $\eta(n) = \eta_i(1 - \frac{n}{n_{max}})$.

obtain the convergence. If the data set is smaller than 10000, s is larger than the one of the previous example.

In the case of the h neighborhood function the best choices of σ_i and σ_f are the following:

$$\sigma_i = \frac{\sqrt{N}}{2} \tag{6.17}$$

$$\sigma_f = 0.01 \tag{6.18}$$

where N is the number of weights.

We have more than one choice for the parameters to obtain the convergence but different choices give different outputs. We illustrate some examples. Finding out 10 weights for a data set of 10000 uniformly distributed data, using $\eta(n) = \frac{\sqrt{6*log(n)}}{\sqrt{(n)}+1}$ and using h if we choose $\sigma_i = 20$ and $\sigma_f = 0.01$ the algorithm converges and the range of values of weights is $(0.48, 0.50)$, in this case the network identifies 10 different values inside that interval; instead if we choose $\sigma_i = 5$ and $\sigma_f = 0.01$ the range is $(0.37, 0.65)$. We see that the best solution is given by (6.17), (6.18), because we have the biggest range of the weights values, in this case is 0.19 to 0.82. It is important to have the range that covers all the interval of the data set because otherwise we do not find the optimal partition. Since we know that for any data distribution the expectation of weights converges, a small range indicates that the network is able to find only some of the limit values of weights, in fact for the reported example, with $N = 10$, we know from the paper of Feng and Tirozzi [Tirozzi (1997)] that the limit values are: $0.05, 0.15, 0.25, 0.35, 0.45, 0.55, 0.65, 0.75, 0.85, 0.95$, so the range of the weights values must be around $(0.05, 0.95)$. In the worst choice the network identifies only one limit value. It happens because the range of action is too large; in this case the algorithm updates simultaneously too many weights and they converge to the same value.

An analogous situation happens using $\Lambda(i, v)$ as a neighborhood function. Using $\eta(n) = \frac{\sqrt{6*log(n)}}{\sqrt{(n)}+1}$, searching always 10 weights for a data set of 10000 uniformly distributed data, if we choose $s \geq 1$ the algorithm converges but the range of weights values change for different choices of s. Increasing s the range of weights becomes smaller and the weights converge to the same limit if $s = 10$. To be more precise if s is equal to N, the network generates N weights (in this case $N = 10$) with the same value. The biggest range, in this case, is obtained with $s = 1$. If the number of weights increases the best choice of s is always the minimum values of s by which we obtain the convergence of the algorithm. For example in the case we search 50 weights the best choice is $s = 3$.

Summarizing to obtain the convergence we must choose $\eta(n)$ with a convenient monotone decay and with a large range; in addition we must estimate the right parameters of the neighborhood function such that we have convergence and the maximum range for the weights values in order to determine the optimal partition of data set.

As we said previously the error of the expectation of the weights varies for different choices of $\eta(n)$, and for some choices of $\eta(n)$ we have a.e. convergence.

This statement is based on the following analysis: we run the Kohonen algorithm 1000 times for different data sequences. We use at the beginning a set of uniformly distributed data of 4000 elements, then $10000, 20000, 30000, 60000, 120000, 150000$ and 250000. This procedure has been done with all the mentioned $\eta(n)$ and both Λ and h. At the end of algorithm running for each data set we have 1000 cases of weights limit values. The mean value of these cases actually converges to the centers of the optimal partition of the interval $(0, 1)$ for all $\eta(n)$ and for each neighborhood function. In addition the average error of limit weights, with respect to the exact values of the centers, decreases on increasing the number of iterations for $\eta(n) = \frac{\sqrt{6*log(n)}}{\sqrt{(n)}+1}$ and $\eta_i(1 - \frac{n}{n_{max}})$ and any neighborhood function; but using $\Lambda(i, v)$ the error decreases more quickly. Moreover the computing time of the algorithm using h is about 7 times longer than the one using Λ and the accuracy of weights on the boundaries is worse using h.

The weights near the border are not updated symmetrically and so they are shifted inward by an amount of the order of $\frac{1}{2*N}$, where N is the number of weights in the case of $\Lambda(i, v)$ while, using h, the weights which are shifted are 4, two for each boundary.

Now we illustrate some of the quoted results. We give examples to illustrate the error evolution using $\eta(n) = \eta_i(1 - \frac{n}{n_{max}})$ and $\eta(n) = \frac{\sqrt{6*log(n)}}{\sqrt{(n)}+1}$, the case of a.e. convergence.

Tables 6.2 and 6.3 concern the application of the algorithm with 4000, 10000, 20000, 30000, 60000, 120000, 150000, 250000 iterations, which are written in the first column, and using Λ as neighborhood function.

As seen in the tables the error decreases faster using $\eta(n) = \eta_i(1 - \frac{n}{n_{max}})$ and it decreases on increasing the iterations; see Figs. 6.5–6.7. In some of these pictures there are the histograms of the limit weights values obtained running the algorithm 1000 times for different numbers M of iterations using every time a specific $\eta(n)$. The histograms show how, increasing M, only in the case of $\eta(n) = \eta_i(1 - \frac{n}{n_{max}})$ and $\eta(n) = \frac{\sqrt{6*log(n)}}{\sqrt{(n)}+1}$ the variance of the histograms tends to 0, each around a limit value of the weight, as we expect since we have a.e. convergence. There are only the histograms for $\eta(n) = \frac{1}{log(n)}$ as example of the convergence in mean since the other cases are similar.

The velocity of convergence is very slow after 100000 iterations, it needs many iterations only to change one weight nearer to its limit value; so to construct the histograms with ten columns we need a huge data set.

The a.e. convergence is guaranteed in the case of $\eta(n) = \eta_i(1 - \frac{n}{n_{max}})$ and $\eta(n) = \frac{\sqrt{6*log(n)}}{\sqrt{(n)}+1}$ by the monotonically decrease of standard deviation of weights as Figs. 6.8 and 6.9 show.

Similar results are obtained with the normally distributed data but for a special choice of the learning parameter. In fact our theorem does not hold for Gaussian distribution as we already mentioned.

Also in this case we have convergence in mean with all the η and for any neighborhood function; and convergence a.e. for $\eta(n) = \eta_i(1 - \frac{n}{n_{max}})$ and $\eta(n) = \frac{\sqrt{6*log(n)}}{\sqrt{(n)}+1}$.

We show in the following histograms (see Figs. 6.10 and 6.11) only the results for a.e. convergence:

Table 6.2 Evolution of the mean error for each weight in the case of $\eta(n) = \eta_i(1 - \frac{n}{n_{max}})$. err_i means: mean error of weight i.

$\eta(n) = \eta_i(1 - \frac{n}{n_{max}})$	$err1$	$err2$	$err3$	$err4$	$err5$	$err6$	$err7$	$err8$	$err9$	$err10$
4000	0.0562	0.0108	0.0183	0.0132	0.0135	0.0144	0.0139	0.0192	0.0110	0.0559
10000	0.0559	0.0088	0.0183	0.0104	0.0104	0.0107	0.0111	0.0185	0.0089	0.0558
20000	0.0565	0.0083	0.0184	0.0097	0.0089	0.0088	0.0092	0.0179	0.0079	0.0560
30000	0.0556	0.0075	0.0180	0.0091	0.0083	0.0082	0.0086	0.0177	0.0073	0.0557
60000	0.0560	0.0071	0.0181	0.0079	0.0073	0.0071	0.0079	0.0179	0.0070	0.0560
120000	0.0559	0.0065	0.0176	0.0072	0.0060	0.0064	0.0072	0.0180	0.0067	0.0560
150000	0.0560	0.0065	0.0180	0.0070	0.0058	0.0060	0.0071	0.0178	0.0065	0.0560
250000	0.0560	0.0062	0.0178	0.0067	0.0050	0.0054	0.0069	0.0180	0.0064	0.0562

Table 6.3 Evolution of the mean error for each weight in the case of $\eta(n) = \frac{\sqrt{6*log(n)}}{\sqrt{n}+1}$. err_i means: mean error of weight i.

$\eta(n) = \frac{\sqrt{6*log(n)}}{\sqrt{n}+1}$	$err1$	$err2$	$err3$	$err4$	$err5$	$err6$	$err7$	$err8$	$err9$	$err10$
4000	0.0559	0.0195	0.0233	0.0203	0.0207	0.0210	0.0211	0.0235	0.0198	0.0556
10000	0.0559	0.0160	0.0216	0.0158	0.0168	0.0161	0.0167	0.0216	0.0158	0.0557
20000	0.0558	0.0139	0.0204	0.0140	0.0142	0.0140	0.0153	0.0195	0.0131	0.0554
30000	0.0558	0.0128	0.0190	0.0135	0.0133	0.0128	0.0136	0.0193	0.0129	0.0562
60000	0.0563	0.0114	0.0189	0.0117	0.0113	0.0114	0.0115	0.0183	0.0108	0.0556
120000	0.0559	0.0098	0.0177	0.0098	0.0096	0.0098	0.0104	0.0184	0.0100	0.0561
150000	0.0559	0.0096	0.0182	0.0096	0.0086	0.0088	0.0093	0.0179	0.0089	0.0557
250000	0.0560	0.0084	0.0179	0.0092	0.0078	0.0081	0.0092	0.0184	0.0089	0.0562

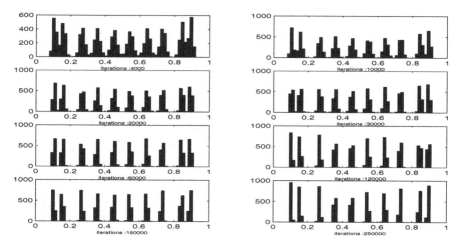

Fig. 6.5   Histograms in the case of $\eta(n) = \eta_i(1 - \frac{n}{n_{max}})$, uniformly distributed data and for different numbers M of iterations.

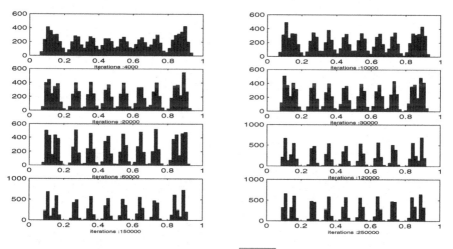

Fig. 6.6   Histograms in the case of $\eta(n) = \frac{\sqrt{6*log(n)}}{\sqrt{(n)}+1}$, uniformly distributed data and for different numbers M of iterations.

Before explaining our application of Kohonen algorithm to microarrays data, we make some remarks on the repetitions of data set presented to the network. This procedure is necessary because the data set is small for microarray data. The accuracy improves by increasing the number of samples and this technique does not change the limit if there is the almost

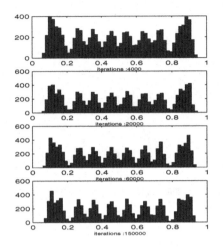 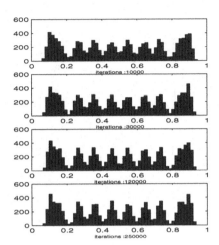

Fig. 6.7   Histograms in the case of $\eta(n) = \frac{1}{log(n)}$, uniformly distributed data and for different numbers M of iterations.

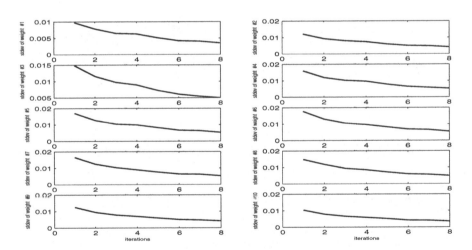

Fig. 6.8   The standard deviation of the weights in the case of $\eta(n) = \eta_i(1 - \frac{n}{n_{max}})$ and uniformly distributed data. The numbers on the x axes indicate the following iterations: 4000, 10000, 20000, 30000, 60000, 120000, 150000, 250000.

everywhere convergence.

To be sure about these findings we have done the same analysis shown above with a data set of 2000 elements repeated at the beginning 2 times,

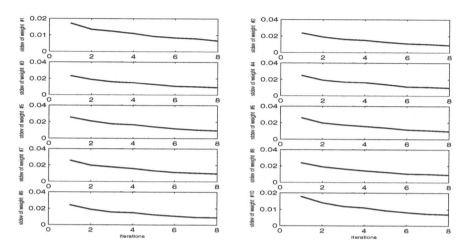

Fig. 6.9 The standard deviation of the weights in the case of $\eta(n) = \frac{\sqrt{6*log(n)}}{\sqrt{(n)}+1}$ and uniformly distributed data. The numbers on the x axes indicate the following iterations: 4000, 10000, 20000, 30000, 60000, 120000, 150000, 250000.

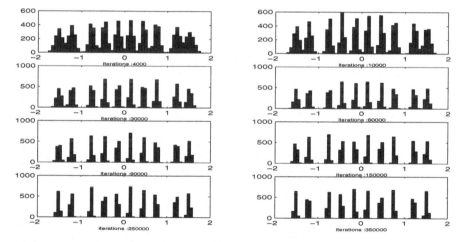

Fig. 6.10 Histograms in the case of $\eta(n) = \eta_i(1 - \frac{n}{n_{max}})$ and normally distributed data and for different numbers M of iterations.

then 5, 10, 15, 30, 60 and 125 so to have the same iterations of the previous analysis. The results are similar, that is we have a.e. convergence for the previous case of $\eta(n)$, that is :

- $\eta(n) = \eta_i(1 - \frac{n}{n_{max}})$

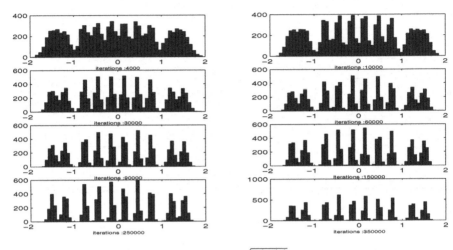

Fig. 6.11 Histograms in the case of $\eta(n) = \frac{\sqrt{6*log(n)}}{\sqrt{(n)}+1}$, normally distributed data and for different numbers M of iterations.

- $\eta(n) = \frac{\sqrt{6*log(n)}}{\sqrt{(n)}+1}$

We show the histograms in Figs. 6.12 and 6.13 in these two cases to illustrate this statement:

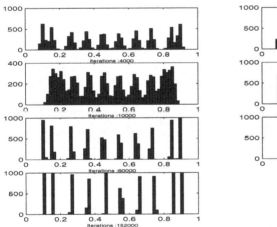
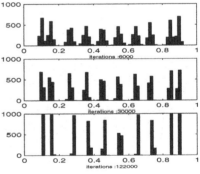

Fig. 6.12 Histograms in the case of $\eta(n) = \eta_i(1 - \frac{n}{n_{max}})$ and uniformly distributed data and for different numbers M of iterations obtained repeating the data several times.

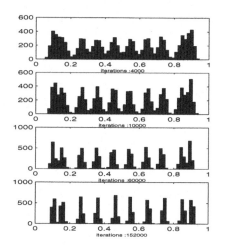
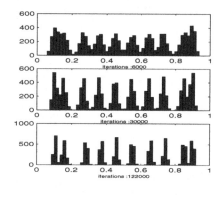

Fig. 6.13   Histograms in the case of $\eta(n) = \frac{\sqrt{6*log(n)}}{\sqrt{(n)+1}}$ and uniformly distributed data and for different numbers M of iterations obtained repeating the data several times.

Application to microarrays data

In this section we show how we have applied the Kohonen network to micro-arrays data set. Following our strategy we have made the cluster analysis of the data for each sample and compared the genes appearing in the nearby clusters, in this way we exploit the neat convergence properties of the one-dimensional case. The set of data we analyzed is the same as the one published in [Quaglino (2004)] where there is an exhaustive description of microarrays sample preparation. In brief total RNA (ttlRNA) was extracted and purified from mammary glands in control and transgenic mice. ttlRNA were pooled to obtain three replicates for the mammary glands of 2-week-pregnant WT BALB/c mice (wk2prg), of 22-week-old un-treated BALB-neuT mice (wk22nt), and of 22-week-old primed and boosted BALB-neuT mice (wk22pb) and two replicates for the mammary glands of 10-week-old untreated BALB-neuT mice (wk10nt). Chips were scanned to generate digitized image data (DAT) files.

DAT files were analyzed by MAS 5.0 to generate background-normalized image data (CEL files). Probe set intensities were obtained by means of the robust multiarray analysis method [Irizarry (2003)]. The full data set was normalized according to the invariant set method. The full-shaped procedure described by Saviozzi et al. [Saviozzi (2003)] was then applied. The resulting 5482 probe sets were analyzed by combining two statistical approaches implemented in significance analysis of micro-arrays [Tusher

(2001)]: two classes unpaired sample method and the multi-classes response test. This analysis produced a total of 2179 probe sets differentially expressed in at least one of the three experimental groups. The 2179 probe sets were converted by virtual of two dye experiments comparing all replicates of each experimental groups with index $j = 1, ..., 3$ (i.e $\frac{wk10nt_i}{wk2prg_j}$ $i = 1, 2$; $\frac{wk22nt_i}{wk2prg_j}$; $\frac{wk22pb_i}{wk2prg_j}$ $i = 1, ..., 3$). Therefore we have 3 replicates of 8 experimental groups.

We apply the Kohonen algorithm to the 2179 probe sets. Our main aim is to detect which genes are up-modulated in wk22pb with respect to wk22nt and wk10nt.

The first step is to implement the one-dimensional Kohonen algorithm in Matlab and study its convergence putting inside as inputs data the 2179 expression levels of genes of any experimental group. In particular our set contains the log values of expression levels of genes, which are normally distributed in any experimental group; so, with regard to the numerical studies done, we choose $\eta_i(1 - \frac{n}{n_{max}})$, with $\eta_i = 0.8$ because we have the almost everywhere convergence, and Λ as neighborhood function, since we have the best accuracy in this case.

For each experimental group the input set Ω is only of 2179 elements so to improve the accuracy we repeat the data presentation set several times in different order. We present the data set 100 times, in such a way that the input set is almost 220000 patterns; in this case the mean error of weights is about 0.01 (as we have seen in our previous studies) and since the average variability of the expression levels of genes among the replicates is about 0.133, this error is acceptable.

We run the Kohonen algorithm fixing N, the number of weights, equal to 30 and then we choose among the limit values found only those weights with a distance greater than two times the average variability of the expression levels of genes among the replicates. We select the weights in this way because otherwise the assignment of a gene to a particular cluster could not be unique. The choice of $N = 30$ has been done analyzing the distribution of the data and considering the variability of the expression levels of genes among the replicates. To obtain weights with a distance greater than two times the average variability of genes expressions we can also fix $N = 15$, but in this way we lose precision in finding weights at the boundaries of the data set interval. It happens because the data are normally distributed, therefore they are concentrated near the mean of the data set and the more we move away from mean the more the distance between weights increases, therefore it is better to choose more weights than those which have an optimal distance between them, such that it is possible to detect more weights at boundaries, since we want to find out up-modulated genes.

Once we have found the limit weights values we separate the data into the identified clusters. This procedure has been done for every experimental group indexed by j ($\frac{wk10nt_i}{wk2prg_j}$ $i = 1, 2$; $\frac{wk22nt_i}{wk2prg_i}$, $\frac{wk22pb_i}{wk2prg_i}$ $i = 1, ..., 3$), so we have eight classifications for each replicate. In addition we choose one of the 24 (8 for each replicate) sequences of limit weights and we separate the data of every experimental into the clusters identified by the chosen sequence. In this way we obtained 24 classifications for every sequence of limit weights (that are 24).

Once we have obtained these classifications we improve the precision of assignment of genes considering their biological variability; therefore we have checked if the expression level of genes which lay on the boundaries of a cluster can be considered really belonging to that cluster or, because its variability, to its neighbor. In detail, if the expression of the genes, incremented of its biological error, is closer to the weight of its cluster than to its nearest weight, the assignment of the gene does not change, otherwise the gene is assigned to the cluster corresponding to its nearest weight.

We can observe that, since the limit weights are ordered, the clusters, with which they are associated, can be ordered in ascending way. Therefore in clusters related to high index we find genes with a greater expression level than in clusters with low index.

For each replicate we select only those genes that are in clusters with high index for the classifications obtained with respect to the limit weights found by analyzing the data of $\frac{wk22pb_i}{wk2prg_i}$ $i = 1, ..., 3$ and in low clusters for classifications obtained by means of the limit weights found by analyzing the data of $\frac{wk10nt_j}{wk2prg_i}$; $\frac{wk22nt_i}{wk2prg_i}$, $j = 1, 2$. After this procedure we have identified a set of 70 up-modulated genes in wk22pb with respect to wk22nt and wk10nt. Among these genes there are 25 that have not been found by Quaglino et al. [Quaglino (2004)]. These new genes found are shown in Tables 6.4 and 6.5.

6.10   Conclusion

We have improved the theorem on the a.e. convergence of the Kohonen algorithm because we prove the sufficiency of a slow decay of the learning parameter, $\sum \eta(n) = \infty$, similar to the one used in the applications. The

Table 6.4   The up modulated genes found out.

Affymetrix ID	Gene Title	Gene Symbol
100376_f_at	similar to immunoglobulin heavy chain	LOC432710
101720_f_at	immunoglobulin kappa chain variable 8 (V8)	Igk-V8
101743_f_at	immunoglobulin heavy chain 1a (serum IgG2a)	Igh-1a
101751_f_at	gene model 194, (NCBI)	Gm194
	gene model 189, (NCBI)	Gm189
	gene model 192, (NCBI)	Gm192
	gene model 1068, (NCBI)	Gm1068
	gene model 1069, (NCBI)	Gm1069
	gene model 1418,(NCBI)	Gm1418
	gene model 1419, (NCBI)	Gm1419
	gene model 1499, (NCBI)	Gm1499
	gene model 1502, (NCBI)	Gm1502
	gene model 1524, (NCBI)	Gm1524
	gene model 1530, (NCBI)	Gm1530
	similar to immunoglobulin light	LOC434586
	chain variable region	LOC545848
	immunoglobulin light chain variable region	ad4
	chain variable region gene model 1420, (NCBI)	Gm1420
102722_g_at	expressed sequence AI324046	AI324046
103990_at	FBJ osteosarcoma oncogene B	Fosb
104638_at	ADP-ribosyltransferase 1	Art1
160927_at	angiotensin I converting enzyme (peptidyl-dipeptidase A) 1	Ace
161650_at	secretory leukocyte peptidase inhibitor	Slpi
162286_r_at	Fc fragment of IgG binding protein	Fcgbp
92737_at	interferon regulatory factor 4	Irf4
92858_at	secretory leukocyte peptidase inhibitor	Slpi
93527_at	Kruppel-like factor 9	Klf9
94442_s_at	G-protein signalling modulator 3 (AGS3-like, C. elegans)	Gpsm3
94725_f_at	similar to immunoglobulin light chain variable region	LOC434033
96144_at	inhibitor of DNA binding 4	Id4
96963_s_at	immunoglobulin light chain variable region	8-30
96975_at	immunoglobulin kappa chain	Igk-V1
	variable 1 (V1)	IgM
	Ig kappa chain	Igk-V5
	immunoglobulin kappa chain	bl1
	variable 5 (V5family)	
	immunoglobulin light chain variable	
	region	
97402_at	indolethylamine N-methyltransferase	Inmt

theorem is not complete because we are not able to prove the necessity of such condition and future work should be concentrated on this point. But for doing such a research one has to find something functional more similar to the Lyapunov functional than the one currently available. This could

Table 6.5   The up modulated genes found out.

Affymetrix ID	Gene Title	Gene Symbol
97826_at	Fc fragment of IgG binding protein	Fcgbp
98452_at	FMS-like tyrosine kinase 1	Flt1
98765_f_at	similar to immunoglobulin heavy	LOC382653
	chain	LOC544903
	similar to immunoglobulin mu-chain	
99850_at	Immunoglobulin epsilon heavy chain constant region	
102156_f_at	immunoglobulin kappa chain	
	constant region	
	mmunoglobulin kappa chain	
	variable 21 (V21)	
	immunoglobulin kappa chain	
	similar to anti-glycoprotein-B of	
	human Cytomegalovirus immunoglobulin Vl chain	
	immunoglobulin kappa chain	
	variable 8 (V8)	
	similar to anti-PRSV coat protein	
	monoclonal antibody PRSV-L 3-8	
	immunoglobulin light chain variable	
	region	
98475_at	matrilin 2	Matn2

make it possible using some argument of convergence similar to the one used for the simulated annealing. We also made many numerical simulations of convergence in order to find the choice of $\eta(n)$ which minimizes the rate of decrease of the average error and also for finding which version of the learning algorithm is better to use. We found that the optimal choice is:

$$\frac{\sqrt{6 * log(n)}}{\sqrt{(n)} + 1} \leq \eta(n) \leq \eta_i(1 - \frac{n}{n_{max}}) \tag{6.19}$$

The algorithm with Λ neighborhood function is better than the one using the function h since it has bad convergence properties. The latter is commonly used in the simulations. After this detailed analysis we applied our optimal choice to the genetic expression levels of tumor cells. The 25 genes identified by us were also consistent with the biological events investigated by Quaglino [Quaglino (2004)].

6.11   Comments and comparison with other algorithms

In this chapter we have dealt with the problem of clustering of gene expression data. This field is very complex, due to the existence of several clustering algorithms which partition the data set from different points of views and tend to emphasize different types of regularities in the data. As a consequence, different clustering algorithms or even the same algorithm along with different input parameters may yield very different partitions given the same data set. This happens because an optimal clustering algorithm does not exist now. Such algorithm should satisfy the following properties:

(1) Scalability and efficiency: algorithms should be efficient and scalable considering the large amount of data to be handled.
(2) Irregular shape: algorithms need to be able to identify a dense set of points which forms a cloud of irregular shapes as a cluster.
(3) Robustness: the clustering mechanisms should be robust against large amounts of noise and outlier.
(4) Order insensitivity: algorithms should not be sensitive to the order of input. That is, clustering results should be independent of data order.
(5) Cluster number: the number of clusters inside the data set needs to be determined by the algorithm itself and not prescribed by the user.
(6) Parameter estimation: the algorithms should have the ability to estimate the internal parameters from the data set without any a priori knowledge of the data.
(7) Stability: no data object will be classified into different clusters for different runnings of the algorithm.
(8) Convergence.

Since an algorithm satisfying all the mentioned properties does not exist, new algorithms are invented continuously each fitting properties $1-8$ better than previous ones but only for a specific distribution of patterns. We think that it would be better to focus the attention on improving the performance of existing algorithms and compare them with the same data set. In such a way we should have a good estimation on advantages and disadvantages of analyzed algorithms and on their improvements.

A step in this direction has been done with the method showed in this chapter and the works done in two theses. The same data set was clustered focusing the attention on different approaches. One approach consisted in computing the PCA and Isomap, two embedding dimensional algorithms. PCA is extensively applied by biologists to have a first view on how the data are divided in large groups. Since these algorithms are not able to investigate more accurately the data set (i.e. it cannot divide the data set into small groups), other techniques are applied, such as: hierarchical clustering, K-means, SOMs, etc. In one thesis, agglomerative hierarchical algorithms have been used, while in the other different types of hierarchical algorithms, SOMs and SOTAs have been applied. The analysis has been done using the software Tmev [3] in which the main clustering algorithms are implemented and using them in more than one dimension.

After these studies we can observe that PCA and Isomap produce the same results, underlying the robustness and stability of the embedding dimensional algorithms. An additive property is that Isomap is able to recognize non-linear structures inside the data and in this contest nonlinear structures have not been found, because the same clusters as the PCA have been obtained as in Figs. 6.14 and 6.15. The other observations concern the applied algorithms of clustering. Using an average linkage hierarchical clustering and choosing the Euclidean distance the same clusters of Quaglino et al. ([Quaglino (2004)]) have been obtained and in one of the clusters new 49 genes have been found but not all these genes have a significant fold change. It has been noticed that the algorithm is extremely sensible to the choice of the metric, it has a deterministic nature: there is no random choice in the iterations of the algorithm. Another feature is the non-uniqueness of the outputs.

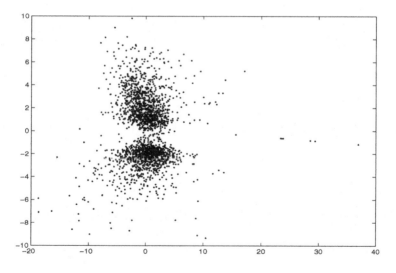

Fig. 6.14   Results obtained by Isomap.

---

[3]Tmev software can be downloaded from: http://www.tm4.org/mev.html

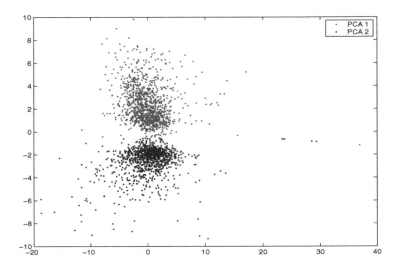

Fig. 6.15   Results obtained by PCA.

The other approach has been to apply several algorithms and to compare them focusing the attention to capture the differences between them. In addition the data set has been partitioned in clusters studying the trends of time courses of every genes. The results are shown in Fig. 6.16. Figure 6.16 shows the trends of the genes found in cluster a found by Quaglino.

In this chapter it has been found that it is better to use the Pearson correlation metric when the time courses are the object of study, while the Euclidean metric is better when it is necessary to study the differences of data values. It has also been shown that hierarchical algorithms suffer the previous mentioned troubles and they are very sensible to the noise of data set. As regard the SOMs produce better results, especially when the choice of learning parameter is of the type (6.19); but they have the drawback of choosing the number of clusters. On the other hand they are very robust to noise of data as all the neural networks. As for SOTA algorithm it is not adapt to divide in few clusters but it improves its qualities when we are searching for more groups.

Summarizing all the results, including those we have showed in this section we can say that:

- PCA is a robust and stable embedding dimensional algorithm but it is not able to divide the data set in small groups.
- Isomap acts as PCA but it gives also information of the existence of

Fig. 6.16   Clusters obtained using time courses of every gene.

non-linear structures inside the data set.

- Hierarchical algorithm is a powerful mean of graphics interpretation and easy to use; but it suffers from the fact that the data objects are grouped initially in clusters chosen almost randomly, with no guarantee of global optimization. This problem is made more complicated by the deterministic nature of hierarchical clustering which implies a critical dependence on the initial choices of the algorithm.

- SOTA is a good compromise between hierarchical and SOMs algorithms but it is not able to divide the data in few clusters and there is no proof of convergence.

- SOM (Kohonen neural network) in one dimension is a powerful mean because, thanks to the property of almost sure convergence for the right choice of learning parameter (as seen in previous sections), is able to identify a unique partition of data set. Therefore it also has

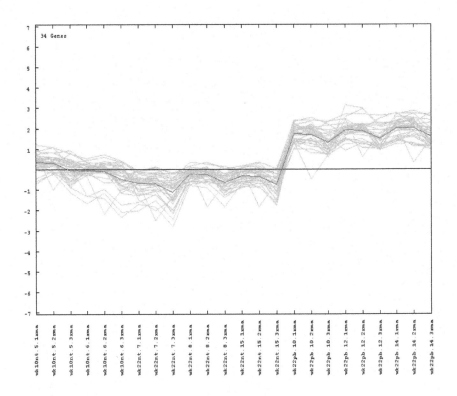

Fig. 6.17   Cluster a found by Quaglino.

the property of order insensitivity, stability and robustness. Its major defect is the choice of number of clusters. Intervention by the users with specific knowledge of the domain can greatly help to enhance the clustering performance.

• SOM in more than one dimension obtains the best results using a learning parameters which satisfies (6.19).

In conclusion we say that the study showed here should only be the beginning of a larger analysis on clustering algorithms including all the existing algorithms and it should be done on the same data set. Meanwhile what the user should do is to know deeply the algorithm which he is applying and the data set. The more information is available the easier the appropriate distance metric and parameters are identified. Then since an optimal algorithm does not exist it is useful to test more algorithms and compare the results, favoring those having rigorous mathematic foundations. If a theorem of almost sure convergence exists this guarantees the best results and it should be privileged; in case there is not such theorem those algorithms defined more rigorously and satisfying the larger number of properties of the optimal algorithm should be chosen.

We have seen that the Kohonen algorithm in one dimension converges almost everywhere for appropriate learning parameters and that makes it powerful and more adaptable; but it has the drawback of the choice of number of clusters. In addition some analysis in one dimension becomes complex. For example the results obtained here are possible only by introducing some additional algorithms which exploit the results of clustering done by Kohonen algorithm. We have the same situation when we want to study the time course of the genes, it is possible to study them but through an additional algorithm. To solve the problem of choice of number of clusters we can apply hierarchical algorithm to carry out the general information and then we can use the Kohonen algorithm. When it is essential to use more than one dimension SOM works well but we have to pay more attention to the outputs because at this moment there exists only a sufficient proof of the convergence of the algorithm.

Chapter 7

Clustering and classification algorithms applied to protein sequences, structures and functions

7.1 Working with proteins information

A protein sustains different levels of biological information: it ranges from the encoded gene sequence that dictates it as a *translation*, to its own three-dimensional shape, which permits it to engage its biological function. Such a high degree of information density is impressive and it naturally reflects the fact that proteins are the main constituents of living matter and are responsible for all the living processes. Protein features are well represented by the four structural levels: primary structure (the amino acids sequence), secondary structure (typical three-dimensional patterns like alpha helix, beta sheets and so on), tertiary structure (the folded shape in the three-dimensional space) and quaternary structure (the multimerization of a protein with copies of itself or with other different proteins in order to assemble a protein complex). In the following we use protein sequence, or simply sequence, to mean amino acid sequence, or primary structure of the protein, while we use protein structure, or simply structure, to mean three-dimensional shape of the protein, or tertiary structure. The four levels are not mutually independent as established by the central dogma of molecular biology, but their correlation is complicated by evolution. Generally speaking, from the nucleotide sequence of a gene we can determine, by the genetic code (see Fig. 7.1), the amino acid sequence of the corresponding protein. Moreover, such primary structure contains all the molecular information needed to build up the spatial shape of the protein, its folded state. Finally, the folded protein is able to accomplish its biological function.

The human genome project and other more recent wide scale experiments aimed at sequencing whole genomes of other organisms produce an overwhelming amount of nucleotide sequences. A small part of these encode for proteins that are still uncharacterized in terms of structure and function. On the other hand, structural genomics (SG) experiments are trying to identify the structure of all the proteins of an organism, as a fundamental effort to give new insight in the identification of the protein function. The complex interrelations among the information that characterize each protein makes the identification of its function a challenge for molecular biologists and bioinformaticians. In this chapter we report and discuss some of the most important technique aimed at classifying proteins' sequence and structure in order to infer their biological function. Finally we describe how, in some cases, clustering techniques can be applied to infer function properties of proteins.

7.2   Protein sequence similarity

The leading approach to the problem of protein function discovery is the analysis of the protein sequence. This implies the comparison of the amino acid sequence of an uncharacterized protein to a database of known proteins, in order to evaluate the similarity of the new protein sequence to

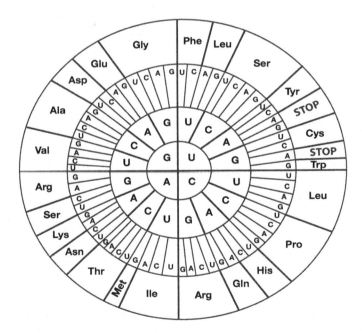

Fig. 7.1   The genetic code: relation between RNA nucleotides triplets and amino acids.

other sequences whose functions are identified. Moreover, sequence analysis allows the construction of phylogenetic trees by which we represent evolutionary relations among proteins. The relationship between two proteins is expressed by the term *homology*, which means that the two proteins have the same phylogenetic origin (i.e. the corresponding genes derived from a common ancestor gene). But sometimes relationships emerge from random events or from convergent evolution at molecular level: a number of genetic mutations could allow two unrelated proteins to share common features. Sequence similarity is a quantitative measure of the relationship between two proteins. It can simply be defined by aligning the two sequences, trying to find as much coincidences as possible between the amino acids of the two proteins (see Tab. 7.1). The straightforward way to discriminate between good and bad alignments is to use a scoring scheme. As a simple example, we could consider the scoring scheme:

(number of matches)−(number of mismatches).

However, the alignment of protein sequences, as that of DNA sequences, requires the definition of a scoring scheme that takes into account the insertion of gaps in the sequences, in order to find better results. Gap insertions are crucial in the alignment of biological sequences since insertions and deletions of amino acids or nucleotides are common events in evolution, giving rise to homologous proteins of different lengths. In the following example, allowing gaps insertion upgrade the alignment to a better one:

```
IPLMTRWDQEQESDFGHKLPIYTREWCTRG
       ⇕⇕⇕⇕⇕⇕⇕⇕⇕⇕
CHKIPLMTRWDQQESDFGHKLPVIYTREW
```

```
IPLMTRWDQEQESDFGHKLP-IYTREWCTRG
⇕⇕⇕⇕⇕⇕⇕⇕⇕ ⇕⇕⇕⇕⇕⇕⇕⇕⇕ ⇕⇕⇕⇕⇕⇕
CHKIPLMTRWDQ-QESDFGHKLPVIYTREW
```

Therefore, a more appropriate scoring scheme for biological sequence alignment must take into account the *opening* of gaps and also their possible extension, if this produces a better alignment of the sequences.

Definition 7.1. Given two sequences A and B, the score S_{AB} associated with the alignment of A and B is defined as:

$$S_{AB} = \sum_{i=1}^{L} s(a_i, b_i) - \sum_{j=1}^{G} \left\{ \gamma + \delta \left[l(j) - 1 \right] \right\} \tag{7.1}$$

Table 7.1   Alignments of amino acid sequences.

Matches	Alignment	Score
1	ALYDFEYG ↕ GLYESEYG	−6
0	ALYDFEYG GLYESEYG	−8
0	ALYDFEYG GLYESEYG	−8
1	ALYDFEYG ↕ GLYESEYG	−6
0	ALYDFEYG GLYESEYG	−8
1	ALYDFEYG ↕ GLYESEYG	−6
0	ALYDFEYG GLYESEYG	−8
5	ALYDFEYG ↕↕ ↕↕↕ GLYESEYG	+2

where $s(a_i, b_i)$ represents the score of the corresponding i-th amino acid pair of the alignment, i ranging from 1 to L, the length of the alignment. G is the number of gaps, while γ and δ represent the gap penalty and the gap extension penalty respectively.

It must be noticed that the single pairing scores $s(a_i, b_i)$ in (7.1) do not measure the identity of the corresponding amino acids in the two aligned sequences. Their value expresses the similarity of the paired amino acids according to a criterion that considers the biochemical properties of the aminoacids [Altschul (1991)] [Altschul (1993)].

Amino acids are not completely independent from a biochemical point of view. Some of them possess common physical or chemical properties, others show similar morphological features (see Fig. 7.2). This biochemical relation implies that casual amino acids mutations could not affect protein function: if a protein carries out a function of particular relevance for a cell to survive, evolution will clearly select only mutations such that (i) they do not affect the functional amino acids or (ii) they replace functional amino acids with others having similar properties while maintain the function of the protein. All other mutations will be counter selected since the cell will not survive. In order to evaluate the similarity between two or more protein sequences, the scoring scheme must then be sensitive of this evolutionary result. The substitution of an aspartic acid (D) with a glutamic acid (E) could be painless for a protein since both of them are hydrophillic (polar) and characterized by an average negative charge (see Fig. 7.2). On the contrary, replacing an aspartic acid with a phenylalanine (F) could affect the entire function of the protein. These two events must be properly evaluated when analyzing protein sequences. The amino acid substitutions are usually evaluated by the so-called substitution matrix [Eidhammer et al. (2004)] [Durbin et al. (1998)] [Ewens and Grant (2001)], a 20×20 symmetric matrix containing 210 distinct values: 20 on the diagonal representing the scores for identical pairing, and 190 off-diagonal values scoring all the possible substitutions. Usually diagonal scores are always positive, while off-diagonal scores could also be negative if the substitution is considered unfavorable. As a trivial example, in the case of the first proposed scoring scheme, where a match exists when the two paired amino acids are identical, the corresponding substitution matrix is the identity

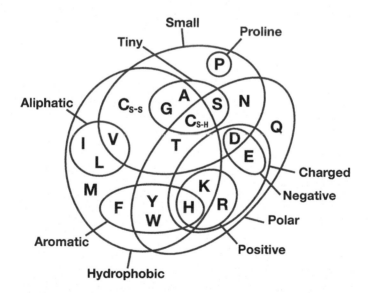

Fig. 7.2 The amino acids properties.

matrix 1I, with all off-diagonal elements equal to 0 and diagonal elements equal to 1. Identity matrix will be adequate if the two sequences to be aligned are very similar, but it will defectively align two distantly related sequences. In general, the evaluation of a transformation rate from the amino acid i into the amino acid j is expressed by log-odds scores:

Definition 7.2.

$$M_{ij} = log_2 \frac{\pi_{ij}}{p_i p_j} \tag{7.2}$$

where π_{ij} represents the empirical frequency with which the substitution is observed, while p_i and p_j are the frequency of the amino acids i and j respectively. Their product represents the expected frequency of substitution, in the hypothesis of amino acids independence. The method by which the score (7.2) is defined depends on the evaluation of the π_{ij}.

More complicated substitution matrices take into account chemical similarity and evolutionary constraints [Altschul (1991)] [Altschul (1993)]. Today it is usual to use matrices based on observed substitution frequencies in families of homologous proteins, like PAM [Dayhoff et al. (1978)] and BLOSUM [Henikoff and Henikoff (1992)] matrices. The sequence alignment algorithm illustrated in Tab. 7.1, in which one sequence *slides* on the other, usually requires a number of comparisons of the order $O(n^2)$, where n is the length of the sequences (supposing that both have the same length). However, if we admit the possibility of gap insertions, the number of operations required by the algorithm grows exponentially, and the algorithm itself become inapplicable. Heuristic methods were then developed, which usually explore only a fraction of the possible alignment, follow complex statistical hypothesis on sequence analysis (for more details see [Durbin et al. (1998)] and [Ewens and Grant (2001)]).

Table 7.2 shows the two fundamental softwares that apply heuristic methods to DNA and protein sequence alignment: BLAST [Altschul et al. (1990)] [McGinnis and Madden (2004)], developed at NCBI (National Center for Biotechnology Information, Bethesda, Maryland, USA) and FASTA [Pearson and Lipman (1989)], developed at EBI (European Bioinformatics Institute, Cambridge, Hinxton, UK). These tools are available on the web, but can also be locally installed for personal applications. Furthermore, the widely used tools ClustalW [Higgins et al. (1994)] and WebLogo [Crooks et al. (2004)] [Schneider and Stephens (1990)] are also reported in Table 7.2. The first is a software tool to perform multiple sequence alignments, in order to identify common motifs in a set of DNA or protein sequences; it

Table 7.2 Sequence alignments Web resources.

Web resource	Address	Function
BLAST	**http://www.ncbi.nlm.nih.gov/BLAST/**	The Basic Local Alignment Search Tool (BLAST) is a tool to identify local similarity regions between sequences. It analyzes nucleotide or protein sequences comparing them to sequence databases and calculating the corresponding statistical significance of matches. BLAST can be used to infer functional and evolutionary relationships between sequences as well as help identify members of gene families.
FASTA	**http://www.ebi.ac.uk/fasta33/**	FASTA is a web tool that provides sequence similarity searching against nucleotide and protein databases by applying the Fasta algorithm. Fasta is very specific when applied to highly diverged sequences for identifying long regions of low similarity. Scanning of complete proteome or genome databases can also be performed.
ClustalW	**http://www.ebi.ac.uk/clustalw/**	ClustalW performs multiple sequence alignment of DNA or proteins. It calculates the best match for the selected sequences, and lines them up so that the identities, similarities and differences can be seen. ClustalW allows the representation of evolutionary relationships by Cladograms or Phylograms.
WebLogo	**http://weblogo.berkeley.edu/logo.cgi**	WebLogo is a web tool designed to make the generation of sequence logos. Sequence logos are a graphical representation of an amino acid or nucleic acid multiple sequence alignment. Each logo consists of stacks of symbols, one stack for each position in the sequence. The overall height of the stack indicates the sequence conservation at that position, while the height of symbols within the stack indicates the relative frequency of each amino or nucleic acid at that position.

also permits the evaluation of cladogram to represent evolutionary relationships among the sequences aligned. The second tool use multiple sequence alignments to produce an effective representation that enhances conserved positions with respect to variable positions.

All the tools reported in Table 7.2 are usually applied to investigate the properties of uncharacterized biological sequences.

7.3 Protein structure similarity

The analysis of protein structure is the main topic of the so-called *structural bioinformatics* and is related with structural data available from X-Ray christallography and NMR experiments and currently stored in public databases like the RCSB PDB (Research Collaboratory Structural Bioinformatics Protein Data Bank, http://www.pdb.org) [Berman et al. (2000)]. It is important to remember briefly that structural bioinformatics involves some special items and challenges that are either not present or not dominant in other types of bioinformatics domain [Altman and Dugan (2003)]:

- Structural data consist of three-dimensional Cartesian coordinates of atoms and are nonlinear and consequently string-based algorithms are not easily applicable to analyze them. The relationships between atoms are also nonlinear (the forces are not linear), which implies expensive computations or approximations.

- For most structural problems, the continuity of the atomic Cartesian coordinates and of the search space implies that algorithms attempting to assign atomic coordinate values have to span infinite spaces.

- There is a fundamental connection between molecular structure and physics, thus limiting the spectrum of approximated representations in order to keep structural calculations physically reasonable.

- Reasoning about structure requires visualization. The development of computer graphics was driven, in part, by the need of structural biologists to look at molecules. However, graphic displays can always be improved such that structural information could be understood by human user but also analyzed by computer programs.

- Structural data, like all biological data, can be noisy and affected by errors. Understanding the protein structural disorder may be critical for understanding the protein's function.

- Protein and nucleic acid structures are generally conserved more than their associated sequence. Thus, sequence will accumulate mutations over time that may make identification of their similarities more difficult, while their structures may remain essentially identical.

In the frame of structural bioinformatics, structure comparisons refers to the analysis of two or more structures looking for similarities in their three-dimensional (3D) structures. To establish equivalences between amino acid residues of two or more protein structures, a procedure of alignment of the protein folds is defined. By *alignment of protein folds* we intend a set of procedures (mainly unitary transformation and error minimization) that try to find the best superposition between the protein structures. More precisely, we must distinguish between pairwise and multiple structural alignment, and between global and local structural alignment:

- Clearly, with the term *pairwise structural alignment* we mean any procedure aimed at superimpose two protein structures in order to find any possible similarity.
- A multiple structure alignment involves the superposition of more than two protein structures. It differs from pairwise alignment also because, as in the case of sequence alignment, it is aimed at detecting conserved structural motifs that characterizes the function of a protein family.
- In contrast with the central dogma, several pairs of proteins having low sequence similarity show a high degree of fold similarity. It is widely accepted that protein structure is much more conserved than protein sequence and seems to be deeply related to protein function and modularity. Therefore it is extremely important to measure the global similarity between two protein structures by a global structural alignment. It is often applied to the main chain of the proteins, neglecting the side chains that have a lower effect on the protein fold.
- A local structural alignment is the result of a procedure aimed at finding local similarities between two non-homologous proteins. Because of convergent evolution events, two proteins unrelated both in sequence and in structure, can show similar disposition of residue side chains concentrated in small portions of the folded structures. This often happens in proximity of the active sites of the two proteins and demonstrates that non-homologous proteins can acquire the same function.

The realization of a structural alignment requires the definition of a score that measures the quality of the alignment. This should depend on the distances among superimposed atoms on the similarity of the corresponding amino acids. The root mean square deviation ($r.m.s.d.$) among atoms

Table 7.3   Structure alignments Web resources.

Web resource	Address
DALI	http://www2.embl-ebi.ac.uk/dali/
SSAP	http://www.biochem.ucl.ac.uk/õrengo/ssap.html
CE	http://cl.sdsc.edu/ce.html
PDBFUN	http://pdbfun.uniroma2.it/
3Dlogo	http://160.80.34.232/3dLOGO/index.html
C-alpha Match	http://bioinfo3d.cs.tau.ac.il/c_alpha_match/
FlexProt	http://bioinfo3d.cs.tau.ac.il/FlexProt/
MultiProt	http://bioinfo3d.cs.tau.ac.il/MultiProt/
MASS	http://bioinfo3d.cs.tau.ac.il/MASS/

represents the widely used score to valuate the quality of a structural alignment.

Definition 7.3. Given the alignment of two protein structures A and B, the root mean square deviation can be calculated as

$$r.m.s.d. = \sqrt{\frac{1}{N} \sum_{i=1}^{N} D_i^2} \tag{7.3}$$

where $D_i = |r_i^A - r_i^B|$ represents the distance between the i-th paired atoms of the protein A and B respectively, and N is the number of atom pairs considered in the alignment.

Therefore the best structural alignment corresponds to a lower $r.m.s.d.$ and a higher number of paired atoms N. There exist several internet resources that perform structural alignments (Table 7.3).

7.4   Protein-protein interaction

Protein-protein interactions play an essential role in the regulation of cell physiology. Not only can the function of a protein be characterized more precisely through its interactions, but also networks of interacting proteins can shed light on the molecular mechanisms of cell life. These interactions form the basis of phenomena such as DNA replication and transcription, metabolic pathways, signalling pathways and cell cycle control. The networks of protein interactions described recently (e.g. [Collins et al. (2007)]; [Gavin et al. (2006)]; [Krogan et al. (2006)]) represent a higher level of proteome organization that goes beyond simple representations of protein networks. They represent a first draft of the molecular integration/regulation of the activities of cellular machineries.

The challenge of mapping protein interactions is vast, and many novel approaches have recently been developed for this task in the fields of molecular biology, proteomics and bioinformatics.

The term *proteomics* was introduced in 1995 [Wasinger et al. (1995)]. This field of research has seen a tremendous growth over the last eleven years, as illustrated by the number of publications related to proteomics. The major goal of proteomics is to make an inventory of all proteins encoded in the genome and to analyze properties such as expression level, post-translational modifications and interactions [Olsen et al. (2006)]. A number of recently described technologies have provided ways to approach these problems. In particular, to identify protein interaction networks, we examine some important experimental techniques in section 7.5.

The final step in the characterization of protein interactions requires the application of bioinformatics tools to process existing experimental information. Bioinformatics tools provide sophisticated methods to answer the questions of biological interest. A number of different bioinformatics strategies have been proposed to predict protein-protein interactions. We describe the more significant methodologies in section 7.6, with a particular attention on the machine-learning approach, which is applied in this work.

Finally, it must be considered that from a computational point of view protein-protein interaction prediction represents a general problem of inference, and bioinformaticians should find the smartest way of using experimental data to obtain successful models, in order to integrate experimental work in the discovery of cellular functioning.

In the following, we will focus on both experimental and computational methods for the study of protein-protein interaction specificity. Specifically, after a brief description of standard experimental techniques and bioinformatical approaches to the problem, we will analyze the development of a neural network model for the prediction of the binders of a given protein domain. We refer to such a methodology as the *neural network approach* to the problem of protein-protein interaction prediction.

7.5 Experimental methods to identify protein ligands

In this section we examine some standard experimental methods to investigate protein-protein interactions. They are *Phage Display, Yeast Two-*

Hybrid, CoImmunoprecipitation, FRET and *Peptide Arrays.*

- **Phage Display**

  The immune system is perhaps the best known natural example of new binding entities detection. When Smith [Smith (1985)] demonstrated that a phage is capable to express on its surface a fusion protein containing a foreign peptide, he also suggested that libraries of fusion phage might be constructed and screened to identify proteins that bind to a specific antibody. So far, there have been numerous developments in this technology to make it applicable to a variety of protein-protein and protein-peptide interaction.

  Usually, the Phage display technique is applied on libraries of random oligopeptides in order to verify their interaction with a protein of interest. The libraries can then be screened to identify specific phage that display any sequence for which there is a binding partner. This screening is performed by a series of affinity purifications known as *panning*. The phage are bound to the partner, which is immobilized on a plastic dish. Phage that do not bind are washed away, and bound phage are eluted and used to infect *E. coli*. Each cycle results in a 1,000-fold or greater enrichment of specific phage, such that after a few rounds, DNA sequencing of the tight-binding phage reveals only a small number of sequences (Fig. 7.3).

- **Two-Hybrid**

  The two-hybrid system is a genetic method that uses transcriptional activity as a measure of protein-protein interaction. It relies on the modular nature of many site-specific transcriptional activators, which consist of a DNA-binding domain and a transcriptional activation domain. The DNA-binding domain serves to target the activator to the specific genes that will be expressed, and the activation domain contacts other proteins of the transcriptional apparatus to enable transcription to occur. Since the two domains of the activator need not be covalently linked, they can be brought together by the interaction of any two proteins. Hence, the system requires the construction of two hybrids: a DNA-binding domain fused to some protein X, and a transcription activation domain fused to some protein Y. If the X and Y proteins interact, they create a functional activator by bridging the activator domain into close proximity with the DNA-binding domain. This can be revealed by the expression of some reporter genes. The assay has been generally performed in yeast cells, but it works similarly

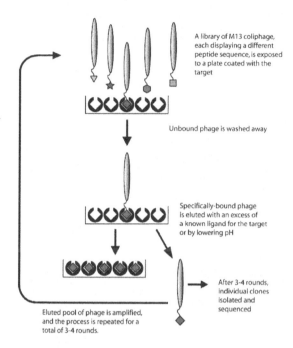

A library of M13 coliphage, each displaying a different peptide sequence, is exposed to a plate coated with the target

Unbound phage is washed away

Specifically-bound phage is eluted with an excess of a known ligand for the target or by lowering pH

After 3-4 rounds, individual clones isolated and sequenced

Eluted pool of phage is amplified, and the process is repeated for a total of 3-4 rounds.

Fig. 7.3   Phage display technique can be extremely effective to identify binding peptides (figure inspired from http://www.chem.tue.nl/).

in mammalian cells and should be applicable to any other eukaryotic cells.

- **Coimmunoprecipitation**
  Coimmunoprecipitation is a classical in vivo method of detecting protein-protein interactions and has been used in literally thousands of experiments. Its basic version involves the generation of cell lysates and the addition of an antibody. Then the antigen is precipitated and washed, and bound proteins are eluted and analyzed.

- **FRET**
  Fluorescence resonance energy transfer (FRET) describes an energy transfer mechanism between two fluorescent molecules. A fluorescent donor is excited at its specific fluorescence excitation wavelength. By a long-range dipole-dipole coupling mechanism, this excited state is then transferred to a second molecule, the acceptor, without photon emissions. The donor returns to the electronic ground state. The described energy transfer mechanism should be more correctly termed *Förster resonance energy transfer*, named after the German scientist Theodor

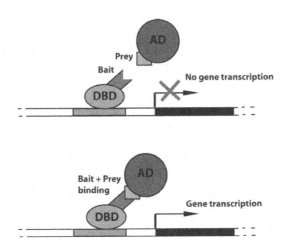

Fig. 7.4 Two-Hybrid system. The bait and the prey are fused to the DNA binding domain (DBD) and to an activation domain (AD) of a transcription factor. If an interaction is established between the two hybrid proteins the transcription of a reporter gene is activated.

Förster, who discovered it. When both molecules are fluorescent, the term *fluorescence resonance energy transfer* is often used, although the energy is not actually transferred by fluorescence.

- **Peptide Arrays**

  The immobilization of peptides on physical supports like cellulose or glass membrane such that it is possible to organize the peptides in high density arrays ($200 peptides/cm^2$), represents a novel technology aimed at studying the whole spectrum of interactors of a given protein module (domain). There are two ways to construct peptide arrays: In the first approach the chemical synthesis of the peptide take place directly on the chosen support (SPOT synthesis), while in the second the peptides are required to be pre-synthesized and then immobilized on the support [Schutkowski et al. (2005)].

7.6   Computational methods to characterize protein ligands

- **Regular expressions**

  The regular expression is a simple way to represent biologically relevant sites common to a family of protein sequences. These sites can be enzyme catalytic sites or regions involved in binding a molecule,

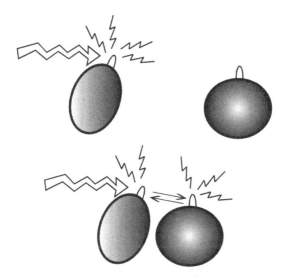

Fig. 7.5   In FRET experiments proteins are tagged with fluorophore molecules. If the interaction between the two proteins is established the fluorophores enter in resonance and the transfer of photons can be detected.

such as ADP/ATP, calcium, DNA and others. Regular expressions can be obtained from multiple sequence alignments, deriving a motif that collects information on conserved residues in specific positions. Several algorithms have been developed to use such regular expression to scan a protein sequence to find a region of match: PatMatch to search for patterns or motifs within all the yeast protein or DNA sequences ($http : //genome - www.stanford.edu/Saccharomyces/help/nph - patmatch.html$), PATTINPROT to scan a protein database with a pattern ($http : //npsa - pbil.ibcp.fr/cgi - bin/npsa\_automat.pl?page = npsa\_pattinprot.html$), ScanProsite [Gattiker et al. (2002)] to scan a protein sequence for the occurrence of patterns and profiles stored in the PROSITE database ($http : //www.expasy.ch/prosite/$), or to search protein databases with a user-entered pattern.

- **Position weight matrices**

  Given a protein and the ligand sequences bound to it, a matrix containing the residue frequencies of occurrence in each position of the ligand can be derived. Such a method defines a position weight matrix (PWM), characterized by twenty rows corresponding to the number of possible amino acids and by a number of columns equals to the meaningful positions of the aligned ligand sequences. Each element within the matrix identified by the row i and the column j, represents the score associated with the amino acid i in the ligand position j, depending on how much the amino acid i appears in the position j of the ligand sequences. PWM can be used to scan and score a protein sequence to find the region with the highest probability of interaction with the given protein. For this scope, PWM assumes independence between position in the sequence, as it calculates scores at each position independently from the amino acids at the other positions. The score of a sequence aligned with a PWM can be interpreted as the log-likelihood of the sequence under a product multinomial distribution. Alternatively, instead of using log-likelihood values in the PWM, several methods use log-odds scores in the PWM. An element in the matrix is then calculated as $m_{ij} = \log(p_{ij}/b_i)$, where p_{ij} is the probability of observing amino acid i at position j of the motif and b_i is the probability of observing the amino acid i in a background model [Ewens and Grant (2001)]. The background model can also be constructed as a completely random model, where each amino acid appears with a uniform probability.

- **Protein docking**

  Molecular docking tries to predict the structure of the intermolecular complex formed between two or more constituent molecules. In particular, the goal of all docking algorithms is to predict with a good reliability the structure of protein-protein or protein-ligand complexes starting from the X-RAY or NMR structures of single molecules, implies the simulation *in silicio* of a protein complex formation procedure starting from the three-dimensional structures of the protein partners. The classical methods that underlie docking investigate geometrical and physico-chemical complementarities between the input structures considered as rigid bodies. However, in the last few years, protein-ligand docking has evolved to a level where full or at least partial flexibility on the ligand is commonly employed.

  Docking protocols can be described as a combination of a search algorithm and a scoring function [Smith and Sternberg (2002)]. The search algorithm should allow the degrees of freedom of the protein-ligand system to be sampled sufficiently as to include the true binding modes. Naturally, the two critical elements in a search algorithm are speed and efficiency in covering the relevant conformational space. Among other requirements, the scoring function should represent the thermodynamics of interaction of the protein-ligand system adequately as to distinguish the true binding modes from all the others explored, and to rank them accordingly.

- **Sequence Prediction of Target (SPOT)**

  The Sequence Prediction Of Target (SPOT) methodology [Brannetti et al. (2000)] [Brannetti and Helmer-Citterich (2003)] consists of a procedure that is designed to infer the peptide binding specificity of any member of a family of protein binding domains. A 20 by 20 matrix is defined for each of the residue pairs that make contact, as derived from the three-dimensional structure of the known complexes. The matrix contains the frequencies with which residue x in the domain is found to make contact with residue y in a peptide ligand. The frequencies are collected from a domain-peptide interaction database obtained by different experimental approaches, such as phage-display and two-hybrids. The ensemble of all the 20×20 matrices, one for each contact position (Fig. 7.6), is represented by a multidimensional array (domain specific matrix) in which the frequency of occurrence of amino acid pairs at each position is stored. The array is then used to provide a score for any putative peptide/domain complex. SPOT can be extended to any protein domain family, if at least one crystal structure of the protein/ligand complex is known and a relatively large database of domain ligand pairs has been assembled.

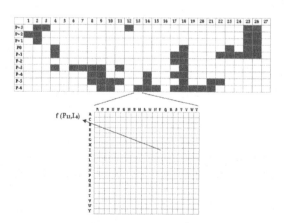

Fig. 7.6   The domain specific matrix of the SPOT method.

7.7   The neural network approach

With the term *neural network approach* we mean all the techniques aimed at develop predicting tools which are based on statistical learning theory, and in particular on neural networks. These applications can be viewed from a statistical point of view as non-parametric models because no distributional hypothesis can be made on their characteristic parameters. Furthermore, such parameters are usually estimated by following an optimization principle and by using complex numerical algorithms. Anyway, the common and basic feature that delineates neural networks is learning from examples. The problem of inducing general functions from specific training examples is the central idea of learning, and learning is the essence of intelligence. If a system could learn and gain from experiences and improve its performance automatically, it would be an advanced tool for solving complex problems such as in the biological systems.

Encoding biological sequence

The application of neural network methodology to biological sequences like nucleic acids or proteins requires proper criteria to transform strings of symbols into a numerical form in order to present the maximum available information to the model. The basic coding assumption so that similar sequences are represented by close vectors and the extraction of the maximal information from the sequence are the requisites that an ideal encoding scheme should have [Wu (1996)]. Actually, the design of the input sequence encoding scheme is application dependent, and can be affected by considerations like the dimension of the encoding sequence window, or the relevance of the different positions within the sequence. Furthermore, it could be important to investigate if the encoding scheme must consider local or global information [Wu (1996)].

Basic encoding methods use binary numbers (0 or 1) arranged in arrays to represent the identity of each molecular residue in the sequence string. For example, a vector of four units with three zeros and a single one is needed for a nucleotide, so the four nucleotides may be represented as 1000 (A), 0100 (T), 0010 (G), and 0001 (C) [Baldi and Brunak (1998)] [Wu (1996)]. Considering protein sequences, a vector of 20 input units (nineteen zeros and a single one) is needed to represent an amino acid [Baldi and Brunak (1998)]. For a sequence of n molecular residues the lengths of these binary representations are $4 \times n$, and $20 \times n$ respectively. Such kind of encodings is usually called orthogonal encodings or sparse encodings and are often considered as the best numerical encodings for categorical variables [Baldi and Brunak (1998)]. The advantages of an orthogonal encoding are its independence from the context of application and the independence of all the possible vectors (in fact they are orthogonal), thus avoiding the inclusion of any spurious correlation or summation property among the n-uples. The disadvantages are that biological symbols could not be independent (there are significant correlations among amino acids) and that the encoding of short protein sequences is sufficient to generate an enormous number of binary variables, thus giving rise to the curse of dimensionality [Bishop (1995)].

A dense representation is also possible, considering only a number of binary variables equals to the integer part of $log_2 n$ (2 for nucleic acids, 5 for proteins). Instead of using the binary numbers to represent the identity of individual sequence residues, real numbers that characterize the residues can also be used. Each residue can be represented by a single feature, such as the hydrophobicity scale, or by multiple properties that may or may not be orthogonal [Schneider and Wrede (1994)]. Each sequence position can also be represented by the residue frequency derived from multiple sequence alignments (i.e. sequence profile of a family; see [Rost and Sander (1993)]).

Data sampling

The organization of the available data from which the machine-learning model can be estimated follows a precise scheme. The complete data set is split into three parts: training set, validation set, and test set. The training set determines the values of the neural network parameters. The validation set is used to determine the stopping time of the training phase. Training continues as long as the performance on the validation set keeps improving. When the improvement ceases, training is stopped. The test set estimates the expected performance (generalization) of the trained network on new data. In particular, the test set should not be used for validation during the training phase. Note that this heuristics requires the application to be data-rich.

Cross-validation is the standard method for evaluating generalization performance with training and test sets. In k-fold cross-validation, the cases are randomly divided into k mutually exclusive test partitions of approximately equal size. The cases not found in each test partition are independently used for training, and the resulting network is tested on the corresponding test partition.

The leave-one-out type of cross-validation is suitable for problems where small sample sizes are available. Here, a model of sample size N is trained using $(N-1)$ cases and tested on the single remaining case, and repeated for each case in turn.

Neural networks paradigm

The training algorithms for a neural network model may be supervised or unsupervised [Wu (1996)]. In supervised training a sequence of training vectors is presented, each with an associated target output vector. Knowing the target output of each training vector corresponds to have an external teacher having knowledge of the environment that is represented by a set of input-output examples. Knowledge of the environment available to the teacher is transferred to the NNs as fully as possible through iterative adjustments to minimize the error signal according to a learning algorithm.

In unsupervised or self-organized learning there is no external teacher to oversee the learning process. Target vectors are unspecified and the learning is normally driven by a similarity measure which acts as a distance between examples. The self-organizing model modifies its parameters so that the most similar vectors (the closest) are assigned to the same output (cluster) unit, which is represented by a centroid vector. Examples of unsupervised training include the Kohonen self-organizing maps [Kohonen (1989)] (see chapter 6) and the adaptive resonance theory (ART) [Carpenter and Grossnerg (1988)].

Neural architecture

A neural network consists of a large number of simple processing elements called neurons [Wu (1996)]. The arrangement of neurons into layers and the connection patterns within and between layers is called the network architecture. Each neuron is connected to other neurons by means of directed communication links, each with an associated weight. The weights represent information being used by the net to solve a problem. Each neuron has an internal state, called its activation level, which is a function of the inputs it has received. An activation function is used to map any real input into a usually bounded range, often 0 to 1 or −1 to 1.

In *feed-forward* (FF) neural network, the information flow from the input units to the output units, in a forward direction: the input units receive the stimulus from the outside world; the output units present the response of the neural network. A multilayer FF network is a net with one or more hidden layers between the input units and the output units (Fig. 7.7). In a fully connected net, every node in each layer of the network is connected to every other node in the adjacent forward layer. If some of the links are missing, we consider the network as partially connected.

Typical feed-forward networks are Perceptron [Rosenblatt (1962)] and Multi-Layer Perceptron (MLP) [Wu (1996)].

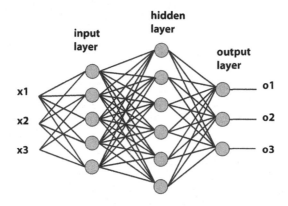

Fig. 7.7   Architecture of a feed-forward multilayer neural network.

The back-propagation (BP) learning rule is fundamental to much current work on neural networks [Chauvin and Rumelhart (1995)]. The generalized delta rule is simply a gradient-descent review of gradient-descent and its many different forms, see expression (7.4). The algorithm provides a computationally efficient method for changing the weights in a feed-forward network, with differentiable activation function units, to learn a training set of input-output examples. BP can be used with a variety of architectures. The elementary BP network is a multilayer perceptron.

$$\Delta w_{ij}(t) = w_{ij}(t + \Delta t) - w_{ij}(t) \propto -\frac{\partial E}{\partial w_{ij}} \qquad (7.4)$$

Here w_{ij} represents the neural weight linking neuron i to neuron j, belonging respectively to two adjacent unit layers in a MLP. The weights are modified during the learning phase accordingly to the minimization of the network's error. E represents the error function.

The BP learning algorithm involves three phases: the feed-forward of the input training pattern; the calculation and back-propagation of the associated error; and the change of the weights in order to minimize the error. In the feed-forward phase, the weights remain unaltered throughout the network, and the function signals of the network are computed on a neuron-by-neuron basis [Wu (1996)]. In the back-propagation phase, error signals are computed recursively for each neuron starting at the output layer, and passed backward through the network, layer by layer (hence, the name *backpropagation*), to derive the error of hidden units. Weights are then adjusted to decrease the difference between the network's output and the target output.

Since the learning here is supervised (i.e. target outputs are available), an error function may be defined to measure the degree of approximation for any setting of the network weights. After training, application of the net involves only the computations of the feed-forward phase. Even if training is slow, a trained net can produce its output very rapidly.

Many enhancements and variations have been proposed for the BP algorithm. These are mostly heuristic modifications with goals of increased speed of convergence, avoidance of local minima, and/or improvement in the network's ability to generalize. A theoretical framework for studying BP was described by [Le Cun (1998)], whose formalism is well suited to the description of many different variations of BP.

PART 3
Appendices

Appendix A

Tutorial of elementary calculus

A.1 Derivation of the results of Chapter 1

The solution of system (1.1) is easily found by integrating the system:

$$\begin{cases} \dfrac{dV}{dt} + \dfrac{V}{\tau} = 0 \\ V(0) = \overline{V} \end{cases}$$

First one has to multiply the above equation by the factor $e^{\frac{t}{\tau}}$, the so-called *integrating factor*, then one remarks that the expression $e^{\frac{t}{\tau}}\left(\dfrac{dV}{dt} + \dfrac{V}{\tau}\right)$ is equal to $\dfrac{d}{dt}e^{\frac{t}{\tau}}V$ by the rule of the derivation of the product of an exponential with a function. Thus we have the following chain of simple identities

$$e^{\frac{t}{\tau}}\left(\frac{dV}{dt} + \frac{V}{\tau}\right) = e^{\frac{t}{\tau}}\frac{dV}{dt} + e^{\frac{t}{\tau}}\frac{V}{\tau}$$

$$= \frac{d}{dt}e^{\frac{t}{\tau}}V(t) = 0 \tag{A.1}$$

Since the first expression of the identities is zero so the last must also be zero. Then this means that the function $e^{\frac{t}{\tau}}V(t)$ is constant and so it must be equal to the value it takes at time $t = 0$

$$e^{\frac{t}{\tau}}V(t) = e^{\frac{0}{\tau}}V(0) = \overline{V} \tag{A.2}$$

and from the last equality one gets immediately that $V(t) = \overline{V}e^{-\frac{t}{\tau}}$. For the solution of the system (A.1)

$$\begin{cases} \dfrac{dV}{dt} + \dfrac{V}{\tau} = -\dfrac{f}{\tau} \\ V(0) = \overline{V} \end{cases}$$

we use the same method as before but there is also a new term on the right-hand side, so the multiplication of both sides by the integrating factor $e^{\frac{t}{\tau}}$ so we get the equations

$$e^{\frac{t}{\tau}}\left(\frac{dV}{dt} + \frac{V}{\tau}\right) = -e^{\frac{t}{\tau}}\frac{f}{\tau}$$

$$= \frac{d}{dt}e^{\frac{t}{\tau}}V(t) = -f\frac{d}{dt}e^{\frac{t}{\tau}} \tag{A.3}$$

The last equation can be solved by integrating both sides from 0 to t so we obtain

$$\int_0^t dt\frac{d}{dt}e^{\frac{t}{\tau}}V(t) = -f\int_0^t dt\frac{d}{dt}e^{\frac{t}{\tau}} \tag{A.4}$$

$$V(t)e^{\frac{t}{\tau}} - V(0)e^{\frac{0}{\tau}} = -f\left(e^{\frac{t}{\tau}} - e^{\frac{0}{\tau}}\right) \tag{A.5}$$

$$V(t)e^{\frac{t}{\tau}} - \overline{V} = -f\left(e^{\frac{t}{\tau}} - 1\right) \tag{A.6}$$

$$V(t) = \overline{V}e^{-\frac{t}{\tau}} - f\left(1 - e^{-\frac{t}{\tau}}\right) \tag{A.7}$$

the last equation is obtained by dividing the factor $e^{\frac{t}{\tau}}$ on both sides.

Appendix B

Complements to Chapter 2

B.1 Solution of the Exercises of Chapter 2

Solutions of Exercise 1.

(a) The rest potential according to the theory of Chapter 1 is $V_{rest} = -f = -60mV$.

(b) The initial condition has no influence on the solution for $t \sim \tau$ so the time is the decay time of the RC circuit $\tau = RC = 10e - 2$ msec.

(c) the change of the potential of the membrane is given by the formula of the capacity of the capacitor $q = C\Delta V$ so $\Delta V = 10e - 1$ volt.

(d) The resistance R should decrease of a factor 10, $R \to R/10$.

Solutions of Exercise 2.

(a)

$$V + f = -\frac{dQ}{dt}R + IR = -CR\frac{dV}{dt} + IR = -\tau\frac{dV}{dt} + IR$$

$$\Downarrow$$

$$\frac{dV}{dt} + \frac{V}{\tau} = -\frac{f}{\tau} + \frac{I}{C} = \frac{IR - f}{\tau}$$

using the formula for the solution of this equation we get $V(t) = (IR - f)(1 - e^{-\frac{t}{\tau}})$.

(b) $IR - f = 1e - 8 \times 1e6 + 70 = 80mv > 60mv$. Thus there is a spike.

(c) $(1e - 8 + \Delta I) \times 1e6 = 1e - 2 + \Delta I \times 1e6 = 60$ mvolt,

$$\Delta I \times 1e6 = 60 - 10 = 50mvolt.$$

$$\Delta I = \frac{50e - 3V}{1e6\Omega} = 50e - 9A = 50nA.$$

(d)

$$T = \tau \log \left[\frac{1}{1 - \frac{\theta}{IR}} \right] = (24e - 4) \log \left[\frac{1}{1 - \frac{60}{70}} \right] = 4.5 \text{msec}$$

(e) Let us solve the equation of the threshold condition for I:

$$IR \left(1 - e^{-\frac{t}{\tau}} \right) = \theta \rightarrow I = \frac{\theta}{R \left(1 - e^{-\frac{t}{\tau}} \right)}$$

putting $T = 9$ nA in the formula one gets $I = 0.96$ nA.

Solutions of Exercise 3.

(a) The solution of the equation of the RC circuit is $V(t) = IR(1 - e^{-\frac{t}{\tau}})$ we have to impose that $V(u) = \theta$, so we have to solve the equation

$$IR \left(1 - e^{-\frac{u}{\tau}} \right) = \theta$$

for u,

$$u = \tau \log \frac{1}{1 - \frac{\theta}{IR}} = 2.2 \quad \text{msec.}$$

(b) $q = C\theta$. C can be found from the value of τ since $\tau = RC$, we get $C = 2.4 \times 10^{-9}$ Farad. Putting this value in the equation for q we get the answer of the exercise.

(c) In this case we can apply simply the first part of the theorem in 2.5. In fact we have that at each arrival of the charge q to the neuron B its potential increases of a quantity

$$\Delta V = \frac{q}{C}$$

with $\frac{q}{C} = \frac{144 \times 10^{-12}}{10^{-9}} = 144$mV. Suppose that we are at the beginning of the process so the neuron B has potential 0. When it receives such a charge, at the times $t_n = nT$ the potential jumps to the value of 144 mV and so it crosses the threshold θ and so it emits a spikes and then the potential goes immediately back to zero. This implies that at the arrival of each spike from neuron A, the neuron B emits a spike so that $t_n = nT$.

(d) It is enough to take I and R of the circuit of the input neuron in such a way that the threshold condition for it is satisfied only once during the intervals $n(u + v), n(u + v) + u$. So it is enough to take $I = 10^{-9}$ A and $R = 10^6$ Ohm, $\theta = 60$ mV. In order to have one spike, periodically with period $T = 4.8$ it is enough to take a neuron with $\tau \sim u/2$, as suggested from

Fig. 2.7, this guarantees an emission of one spike every $u + v$ seconds. So one can choose u and v in such a way that $u + v = 4.8$ msec for example $v = 1.8$ msec and then $u = 3$ msec and so the neuron emitting spikes should have a $\tau \sim 1.5$ msec.

(e) The answer is a straightforward consequence of the previous answer.

Solution of Exercise 4.

The potential satisfies two different equations according to the time interval

$$\begin{cases} \frac{dV}{dt} + \frac{V}{\tau} = I_1 & \text{for} \quad 0 \le t \le u \\ \frac{dV}{dt} + \frac{V}{\tau} = I_2 & \text{for} \quad u \le t \le u + v \\ V(0) = 0 \end{cases}$$

with conditions of continuity at the end of the interval $(0, u)$.

(a) The solution of the first equation is given by

$$V(t) = I_1 R \left(1 - e^{-\frac{t}{\tau}} \right) \quad \text{for} \quad 0 \le t \le u.$$

Thus the spiking condition in the interval $(0, u)$ is that the maximum of the potential in the interval is larger than θ, since the maximum is reached at the time $t = u$ we have:

$$I_1 R \left(1 - e^{-\frac{u}{\tau}} \right) > \theta$$

(b) In the second interval the value reached by the potential V(t), in the absence of spikes in the interval $(0, u)$, at the time u is

$$V(u) = I_1 R \left(1 - e^{-\frac{u}{\tau}} \right)$$

If we integrate the equation in the interval $(u, u + v)$ we obtain

$$V(t) = I_1 R \left(1 - e^{-\frac{u}{\tau}} \right) e^{-\frac{t-u}{\tau}} + I_2 R \left(1 - e^{-\frac{t-u}{\tau}} \right).$$

It is easy to see that the maximum of this function is obtained for $t = u + v$, since its derivative with respect to t is always negative, since $I_1 \left(1 - e^{-\frac{u}{\tau}} \right) e^{-\frac{t-u}{\tau}} < I_2$ as a consequence of the hypothesis that $I_1 < I_2$. If one computes the value of this function at the time $t = u + v$ then the inequality of the exercise is obtained.

Solution of Exercise 5.

Applying the same argument as in the previous exercise we can write the solution:

$$V(t) = I_1 R(1 - e^{-\frac{t}{\tau}}) \quad \text{for} \quad 0 \le t \le u \tag{B.1}$$

$$V(t) = I_1 R(1 - e^{-\frac{u}{\tau}})e^{-\frac{t-u}{\tau}} + I_2 R(1 - e^{-\frac{t-u}{\tau}}) \quad \text{for} \quad u \le t \le u+v \tag{B.2}$$

$$V(t) = e^{-\frac{t-(u+v)}{\tau}} \left[I_1 R(1 - e^{-\frac{u}{\tau}})e^{-\frac{v}{\tau}} + I_2 R(1 - e^{-\frac{v}{\tau}}) \right] \tag{B.3}$$

for $u + v \le t \le u + v + w$. This solution repeats periodically under the hypothesis $w \gg \tau$ since the potential is zero when the new impulse current arrives at the time $u + v + w$. In the hypotheses of the exercise and since $u, v, w \gg \tau$ the maximum value of the potential in the interval $(0, u)$ is $I_1 R$ since the exponent $e^{-\frac{u}{\tau}}$ can be neglected while the maximum value of the potential in the interval $(u, u + v)$ is $I_2 R$ for an analogous reason and so one gets the inequalities (2.9), (2.10).

Solution of Exercise 6.

From the solution of the previous exercise it follows that there is a spike in the interval $(0, u)$ if

$$I_1 R(1 - e^{-\frac{u}{\tau}}) \ge \theta.$$

Moreover if one wants to have only one spike in the interval $(0, u)$ at the time $t_1 \sim u$ then we must impose that

$$t_1 = \tau \log \frac{1}{1 - \frac{\theta}{I_1 R}} \sim u$$

which is the first condition. Since the spike in the interval $(0, u)$ took place at the time $t \sim u$ we have that the solution in the interval $(u, u+v)$ is given by

$$V(t) = I_2 R(1 - e^{-\frac{t}{\tau}})$$

since the initial condition $V(u) \sim 0$ because a spike took place at time $t = u$. Then, if t_2 is the time of the spike in the interval $(u, u + v)$, if we want to have only one spike we need to impose that

$$v - (t_2 - u) < t_2 - u$$

since after the spike the potential is reset to zero and the time necessary for the neuron to emit a spike must be less than the time necessary for arriving at the end of the interval. We get:

$$t_2 > v/2 + u$$

and since

$$t_2 = \tau \log \frac{1}{1 - \frac{\theta}{I_2 R}} + u$$

we get the following condition for the second spike

$$\tau \log \frac{1}{1 - \frac{\theta}{I_2 R}} > v/2.$$

Solution of Exercise 7.

We have to solve the following system of differential equations:

$$\frac{dV_1}{dt} + \frac{V_1}{\tau} = I\frac{t}{C\tau} \quad \text{for} \quad 0 \le t \le u \quad V_1(0) = 0 \tag{B.4}$$

$$\frac{dV_2}{dt} + \frac{V_2}{\tau} = I\frac{u}{C\tau} \quad \text{for} \quad u \le t \le u + v \quad V_1(u) = V_2(u) \tag{B.5}$$

where we named with $V_1(t), V_2(t)$ the solutions in the intervals $(0, u)$ and $(u, u + v)$ respectively. The last condition in equation (B.5) is a condition of continuity of the potential: the two solutions must be equal at the point u where the equation changes form, since the right-hand side is continuous at that point. We follow the usual procedure of multiplication by the integration factor in both the equations of the system:

$$\begin{cases} e^{\frac{t}{\tau}} \left(\frac{dV_1}{dt} + \frac{V_1}{\tau}\right) = e^{\frac{t}{\tau}} I \frac{t}{C\tau} & \text{for} \quad 0 \le t \le u \\ e^{\frac{t}{\tau}} \left(\frac{dV_2}{dt} + \frac{V_2}{\tau}\right) = e^{\frac{t}{\tau}} I \frac{u}{C\tau} & \text{for} \quad u \le t \le u + v \end{cases}$$

where the main difference among the two equations is that the second one is the usual equation with constant r.h.s. while the first has the linear term $e^{\frac{t}{\tau}} I \frac{t}{C\tau}$ which changes the integration. So one gets integrating the equation (B.4) in $(0, t)$

$$e^{\frac{t}{\tau}} V_1(t) - V_1(0) = \frac{I}{C} \int_0^t ds \frac{s}{\tau} e^{\frac{s}{\tau}}$$

$$= \frac{I}{C} \left[se^{\frac{s}{\tau}} \Big|_0^t - \int_0^t ds e^{\frac{s}{\tau}} \right] = \frac{I}{C} \left[te^{\frac{t}{\tau}} - \tau(e^{\frac{t}{\tau}} - 1) \right]$$

dividing the equation by the factor $e^{\frac{t}{\tau}}$ we get

$$V_1(t) = \frac{I}{C}\left(t - RC(1 - e^{-\frac{t}{\tau}})\right) = IR\left(\frac{t}{\tau} - (1 - e^{-\frac{t}{\tau}})\right).$$

The solution in the interval $(u, u + v)$ is obtained with the usual formula

$$V_2(t) = V_2(u)e^{-\frac{t-u}{\tau}} + IR\frac{u}{\tau}(1 - e^{-\frac{t-u}{\tau}})$$

$$= IR\left(\frac{u - t_1}{\tau} - (1 - e^{-\frac{u-t_1}{\tau}})\right)e^{-\frac{t-u}{\tau}} + IR\frac{u}{\tau}(1 - e^{-\frac{t-u}{\tau}}).$$

We subtracted t_1 from u in the first term of the sum because there has been a spike at time t_1 and so the potential start to have the same behavior as it had in the interval $(0, t_1)$ but with the time $t - t_1$, t_1 being the time when the potential is zero. Since the derivative of the function $\frac{t}{\tau} - (1 - e^{-\frac{t}{\tau}})$ is positive for $t \in (0, u)$ and it is equal to zero only for $t = 0$ where the function has a minimum, then the maximum of $V_1(t)$ is attained for $t = u$ thus the condition for having a spike in this interval is that the maximum of the potential in that interval is larger than the threshold i.e.

$$IR\left(\frac{u}{\tau} - (1 - e^{-\frac{u}{\tau}})\right) > \theta.$$

Whereas in order to have the spike in the second interval we have to impose that $max_{t\in(u,u+v)}V_2(t) > \theta$. It is easy to see that the derivative of $V_2(t)$ in the interval $(u, u + v)$ is positive so $V_2(t)$ also is an increasing function. Thus the condition of spiking is $V_2(u + v) > \theta$ which gives the condition of the exercise

$$IR(\frac{u - t_1}{\tau} - (1 - e^{-\frac{u-t_1}{\tau}}))e^{-\frac{v}{\tau}} + IR\frac{u}{\tau}(1 - e^{-\frac{v}{\tau}}) > \theta$$

B.2   Matlab programs

We describe here the Matlab programs which have generated the figures in the text so as to give the reader some training on the use of Matlab and how to realize simulations of neural systems using Matlab. We start with the program which computes the potential $v(t)$ of a neuron described by an Integrate & Fire model with a constant input current I. It computes the function $v(t)$, the interspike time, and plots the graph of the function. It is not necessary to make such a program because the analytic solution of the problem has been easily derived but its generalization to more complex cases is straightforward. This is the Matlab program which generates Fig. 2.2.

```
clear all
close all
pass_int=1000;
delta=10e − 5;
t=zeros(1,pass_int);
v=zeros(1,pass_int);
s=zeros(1,pass_int);
I=60e − 9;
R=1e9;
C=1e − 10;
f=I/C;
tau=R*C;
theta=5;
h=delta/tau;
v(1)=0;
t(1)=delta;
s(1)=theta;
k=1;
t_spike(1)=0;
for i=1:pass_int-1
v(i+1)=v(i)-h*v(i)+f*delta;
if v(i+1)≥ theta
v(i+1)=0;
k=k+1;
t_spike(k)=i*delta;
t_interspike(k)=t_spike(k)-t_spike(k-1);
end
t(i+1)=t(i)+delta;
s(i | 1)=theta;
end
T_spike=tau∗ log(1/(1 − theta/I ∗ R))
plot(1e-3,0.6)
hold on
plot(t,v)
plot(t,s)
```

Let us comment the single commands:

(1) is used for erasing all the values of the variables computed by the program in the previous run.

(2) close all the figures and subroutines still acting.

(3) since we are doing a numerical integration, this variable declares how many steps we are going to use.

(4) delta is the elementary increment of t: $t \to t + \delta$

(5) is a declaration that t (the time) is a vector with zero components and with pass_int components

(6) is a declaration that v (the potential) is a vector with zero components and with pass_int components

(7) is a declaration that s (the threshold) is a vector with zero components and with pass_int components, it is not necessary to have it as a vector but it is necessary for plotting it.

(8) I (Ampere) is the input current which is kept constant.

(9) R (Ohm) is the resistance of the RC circuit of the Integrate & Fire model.

(10) C (Farad) is the capacity of the RC circuit of the Integrate & Fire model.

(11) $f = I/C$ is the term on the r.h.s. of the equation of the Integrate & Fire model.

(12) $\tau = RC$ (seconds) is the decay time of the membrane potential of the Integrate & Fire model.

(13) θ is the value of the threshold of the Integrate & Fire model.

(14) h is the ratio of the time interval δ and τ appearing in the numeric evaluation.

(15) $v(1) = 0$ is the initial value of the potential.

(16) $t(1) = 0$ is the time corresponding to $v(1)$.

(17) $s(1) = 0$ is the value of the threshold which is plotted as a vector together with v and t.

(18) $k = 1$ is the index of the vector t_spike with components equal to the spiking times.

(19) t_spike(1) $= 0$ is the first spike time which is put equal to zero since it corresponds to the origin of the time.

(20) **for** i=1:pass_int-1, is the main operation of the program. It is a loop, i.e. for any fixed integer value of i in the interval (1, pass_int) the operation encountered in the loop are executed and when they are finished $i \to i+1$ and the program starts again from the beginning of the loop.

(21) the method of integration, the derivative $\frac{dv}{dt}$ is approximated with the ratio $\frac{v(i \times \delta + \delta) - v(i \times \delta)}{\delta}$ and the equation obtained substituting this expression for

the derivative in the equation for v is solved with respect to $v((i+1) \times \delta)$ obtaining the expression for $v(i+1)$ written in the command.

(22) this is the condition on the potential, if it is bigger than the threshold it is put equal to zero in the next line.

(23) the potential at time $(i+1) \times \delta$ is set to zero for the emission of the spike.

(24) if a spike took place at the time $i \times \delta$ then the number of spikes, k, must be incremented by 1.

(25) the spiking time $i \times \delta$ becomes the $k+1$ component of the vector of the spiking times (t_spike).

(26) the interspike time, interspike(k), is computed as the difference of the two subsequent spike time t_spike(k)-t_spike($k-1$).

(27) the **end** command signals the end of the operation of the **if** command which controls all the operations of checking and computing the spike times and the interspike.

(28) the new component of the vector of the running times $t(i+1)$ is computed just summing δ to the previous time $t(i)$.

(29) for having also the threshold plotted on the graph one needs also to set $s(i+1) = \delta$.

(30) the formula of the interspike time is used for checking the value of the interspike time computed numerically.

(31) **plot**(1e-3,0.6) in order to have the threshold drawn below the upper line of the graph we plot a point which is taken as the maximum y coordinate by the plot program.

(32) **hold on** makes the lines plotted in the same graph.

(33) **plot**(t,v): plots the times $t(i)$ on the horizontal axis and the corresponding value $v(i)$ on the vertical axis giving the plot of the potential.

(34) **plot**(t,s): plots a horizontal line at the height of the threshold θ in the graph.

The program below generated Fig. 2.5, it solves numerically the problem when the neuron receives a constant input current for u seconds and after the current is zero. The only difference is the insertion of a new block:

```
if   i*delta ≥ u
f1=0;
else
f1=f;
end
```

The commands of this block set the r.h.s. of the model equal $f = I/C$ if $t \in (0, u)$ otherwise it is equal to zero. The rest of the commands are the same. The condition $t \notin (0, u)$ is given by **if** i*delta $\geq$ u since the time is $i\delta$ in the numerical integration.

```
passint=15000;
delta=10e − 5;
u=1;
I=20e-9;
R=1e9;
C=1e-10;
soglia=10;
tau=R*C;
h=delta/tau;
v(1)=0;
t(1)=delta;
s(1)=theta;
k=1;
t_spike(1)=0;
for i=1:pass_int-1
v(i+1)=v(i)-h*v(i)+f*delta;
if   i*delta ≥ u
f1=0;
else
f1=f;
end
if v(i+1)≥ theta
v(i+1)=0;
k=k+1;
t_spike(k)=i*delta;
t_interspike(k)=t_spike(k)-t_spike(k-1);
end
t(i+1)=t(i)+delta;
s(i+1)=soglia;
end
T_spike=tau* log(1/(1 − theta/I * R))
plot(1e-3,0.6)
hold on
plot(t,v)
plot(t,s)
```

The next program is the one which integrates the model with input given by the periodic current of the type Fig. 2.6 and it is the one who created Figs. 2.7–2.9. These figures are obtained by varying the decay rate τ of the neuron. If it is short with respect to u then there is repetitive firing of the neuron inside the interval $(n(u + v), n(u + v) + u)$. If τ increases and is of the order of $u/2$ then there is only one spike in the intervals $(n(u + v), n(u + v) + u)$ but with the same structure, if τ is of the order of some multiples of $u+v$ then there is a spike every few intervals $n(u+v)$. The new block in the middle of the program is done for realizing the periodic structure of the model and then the integration goes on automatically.

```
clear all
close all
delta=10e-3;
U=10;
V=20;
I=1e-9;
R=2e8;
C=1e-7/2;
f=I/C;
tau=R*C;
soglia=I*R/2;
passint=5*(U+V)/delta;
t=zeros(1,pass_int);
v=zeros(1,pass_int);
s=zeros(1,pass_int);
Q1=zeros(1,pass_int);
Q2=zeros(1,pass_int);
Z1=zeros(1,pass_int);
Z2=zeros(1,pass_int);
F=zeros(1,pass_int);
h=delta/tau; v(1)=0;
s(1)=theta;
for n=1:pass_int
t(n)=n*delta;
for k=1:5
Q1(n)=(t(n)-(k-1)*(U+V));
Q2(n)=((k-1)*(U+V)+U-t(n));
Z1(n)=(t(n)-k*U-(k-1)*V);
```

```
Z2(n)=(k*(U+V)-t(n));
Q(n)=Q1(n)*Q2(n);
Z(n)=Z1(n)*Z2(n);
if Q(n)≥ 0
F(n)=f;
end
if Z(n)≥ 0
F(n)=0;
end
end
end
k=1;
s(1)=theta;
t_spike(1)=0;
v(1)=0;
for i=1:pass_int-1
v(i+1)=v(i)-h*v(i)+F(i)*delta;
if v(i+1)> theta
v(i+1)=0;
k=k+1;
t_spike(k)=i*delta;
t_interspike(k)=t_spike(k)-t_spike(k-1);
end
s(i+1)=theta;
t(i+1)=t(i)+delta;
end
figure(1);
plot(0,2*I)
hold on
plot(t,F*C)
hold off
figure(2);
plot(0,I*R*1.2)
hold on
plot(t,F*tau)
plot(t,v)
plot(t,s)
```

Appendix C

Complements to Chapter 3

C.1 Main definitions of matrix calculus

We give here the definition of matrix calculus in the simple case of 2×2 matrices which are used in Chapter 3. Let us start with the general definition of a matrix:

Definition C.1

A 2×2 matrix A is an array with two columns and two rows

$$A = \begin{pmatrix} a_{11} & a_{12} \\ a_{21} & a_{22} \end{pmatrix}$$

the quantities a_{ij} are real or complex numbers and are called the *matrix elements* of A. The determinant of the matrix is a real or complex number $\mathrm{Det} A$ associated to the array with the following rule:

$$\mathrm{Det} A = a_{11}a_{22} - a_{21}a_{12}. \tag{C.1}$$

We need to know also what is the identity matrix I, the equivalent of 1 for the numbers, and also how to make the sum or differences among matrices and the multiplication with a real number α.

Definition C.2

Define the 2×2 identity matrix I as

$$\mathrm{I} = \begin{pmatrix} 1 & 0 \\ 0 & 1 \end{pmatrix}$$

and the sum or difference of matrices as the matrix obtained by subtracting or adding the corresponding matrix elements

$$A \pm B = \begin{pmatrix} a_{11} & a_{12} \\ a_{21} & a_{22} \end{pmatrix} \pm \begin{pmatrix} b_{11} & b_{12} \\ b_{21} & b_{22} \end{pmatrix} = \begin{pmatrix} a_{11} \pm b_{11} & a_{12} + \pm b_{12} \\ a_{21} \pm b_{21} & a_{22} \pm b_{22} \end{pmatrix}$$

We need to define the multiplication of a matrix A with a real number α as the matrix obtained multiplying each matrix element by α:

Definition C.3

$$\alpha A = \begin{pmatrix} \alpha a_{11} & \alpha a_{12} \\ \alpha a_{21} & \alpha a_{22} \end{pmatrix}$$

We also need to define the eigenvalues of a matrix A and the secular equation, these two concepts being the basis of the proof of the Hopf bifurcation and generation of the spikes.

Definition C.4 The *eigenvalues* $\lambda = (\lambda_1, \lambda_2)$ of a 2×2 matrix A are those numbers (real or complex) which satisfy the equation (*the secular equation*):

$$\mathrm{Det}(A - \lambda \mathrm{I}) = 0$$

This equation takes a simple form since the matrix appearing in the above expression is easy to compute:

$$A - \lambda \mathrm{I} = \begin{pmatrix} a_{11} & a_{12} \\ a_{21} & a_{22} \end{pmatrix} - \lambda \begin{pmatrix} 1 & 0 \\ 0 & 1 \end{pmatrix} = \begin{pmatrix} a_{11} - \lambda & a_{12} \\ a_{21} & a_{22} - \lambda \end{pmatrix}$$

and by the definition of the determinant we get the simple expression:

$$(a_{11} - \lambda)(a_{22} - \lambda) - a_{12}a_{21} = 0. \tag{C.3}$$

C.2   Matlab programs for integrating the *FN* and *HH* models

Integration of the FN model, main program:

```
clear all
close all
[T, Y]=ode15s(@fitzhugh,[0 1000], [0,0]);
figure
subplot(2,2,1)
plot(T, Y(:,1))
subplot(2,2,2)
plot(T,Y(:,2))
subplot(2,2,3) plot(Y(:,1),Y(:,2))
```

Note that this program is very simple: all the integration method is in fact contained in the command **ode15s(@fitzhugh,[0 1000], [0,0])**. This is a call to the Matlab subroutine ode15s which contains a good numerical integration method, then the system of equation, the parameters are contained in the Matlab function, to be made by the programmer, which must have the name fitzhugh.m, this name should coincide with the name and function call of the first line of the function fitzhugh.m: **function dy=fitzhugh(t,y)**. Then one must follow the syntax rules used in the function fitzhugh.m. The other parameters of the call $[0, 1000]$ and $[0, 0]$ are respectively the time interval of integration and the initial point. The same rule are applied for the program which integrates the HH model.

The function containing the equations and the parameters of the FN model:

```
function dy=fitzhugh(t,y)
dy=zeros(2,1); a=0.5;
b=1/10;
gamma=1/4;
q=(1 + a)^2-3*(a-b*gamma);
v_c_1=((1+a)-sqrt(q))/3;
v_c_2=((1+a)+sqrt(q))/3;
x=v_c_1;
I_1=x*(x^2-x*(1+a)+a+1/gamma);
x=v_c_2;
I_2=x*(x^2-x*(1+a)+a+1/gamma);
alpha=0.3;
I=I_1+alpha*(I_2-I_1);
dy(1)=y(1)*(1-y(1))*(y(1)-a)-y(2)+I;
dy(2)=b*(y(1)-gamma*y(2));
```

Integration of the HH model, main program:

```
clear all
close all
[T,Y]=ode15s(@hh,[0 100],[-60,0,0,0]);
subplot(2,2,1)
plot(Y(:,1))
subplot(2,2,2)
plot(Y(:,2))
```

```
subplot(2,2,3)
plot(Y(:,3))
subplot(2,2,4)
plot(Y(:,4))
```

The function containing the equations and the parameters of the HH model

function dy=hh(t,y)
dy=zeros(4,1); y(1): Potential
y(2)=m: activaton of Na channel
y(3)=h: disactivation of Na channel
y(4)=n: activation of K
channel C=1;
g_Na=120/C;
g_K=36/C;
g_L=0.3/C;
V_K=-77;
V_Na=50;
V_L=-54.4;
I=10/C;
alpha_m=0.1*(y(1)+40)/(1-exp(-(y(1)+40)/10));
beta_m=4*exp(-(y(1)+65)/18);
tau_m=1/(alpha_m+beta_m);
m_as=alpha_m*tau_m;
alpha_h=0.07*exp(-(y(1)+65)/20);
beta_h=1/(1+exp(-(y(1)+35)/10));
tau_h=1/(alpha_h+beta_h);
h_as=alpha_h*tau_h;
alpha_n=0.01*(y(1)+55)/(1-exp(-(y(1)+55)/10));
beta_n=0.125*exp(-(y(1)+65)/80);
tau_n=1/(alpha_n+beta_n);
n_as=alpha_n*tau_n;
```

**dy=-g_Na*$y(2)^3$*y(3)*(y(1)-V_Na)-g_K*$y(4)^4$*(y(1)-V_K)**
**   -g_L*(y(1)-V_L)+I;**
**dy(2)=(m_as-y(2))/tau_m;**
**dy(3)=(h_as-y(3))/tau_h;**
**dy(4)=(n_as-y(4))/tau_n;**
```

Appendix D

Complements to Chapter 4

D.1 A simple introduction to probability

We give here a simple introduction to the main definitions and concepts of probability in order to make the book self-contained. We think that the better way to introduce these concepts is through the solutions of some elementary problems.

(I) Compute the probability of events of a poker game. These are the simplest events to discuss because they form a discrete set. We consider the case of four players so there are 28 cards (Ace (A), King (K), Queen (Q), Knave (J), 10, 9 and 8) with four different suits (hearts, diamonds, clubs, spades). We suppose that each person get the cards one by one from the pack of cards.

(a) Compute the probability getting three aces in a sequence in the first three extractions.

A. The probability of extracting the first ace is $4/28$ the number of positive cases divided by the number of possible cases. The probability of getting the second ace after the first is $3/27$ because the number of aces decreased by 1 as well as the number of cards, thus the probability of getting the two aces in a row is $(4/28)(3/27)$ and the probability of getting the third is $2/26$. The probability of the event will be the product of three probabilities multiplied by all the possible orders of extraction of the suits $3!$

$$3! \frac{4}{28} \frac{3}{27} \frac{2}{26}.$$

(b) Compute the probability of getting a maximal straight (AKQJ10).

A.

$$5!\frac{4}{28}\frac{4}{27}\frac{4}{26}\frac{4}{25}\frac{4}{24}.$$

(c) Compute the probability of getting a double couple of the type AAKK.
A.

$$4!\frac{4}{28}\frac{3}{27}\frac{4}{26}\frac{3}{25}.$$

(II) Some problems taken from internet. David Morgan, city manager of Yukon, Oklahoma, must negotiate new contracts with both the fire fighters and the police officers. He plans to offer to both groups a 7% wage increase and hold firm. Mr. Morgan feels that there is one chance in three that the fire fighters will strike and one chance in seven that the police will strike. Assume that the events are independent.

(a) What is the probability that both will strike? A. We have that

$$\text{Prob(fire fighter's strike)} = \frac{1}{3}$$

$$\text{Prob(police officer's strike)} = \frac{1}{7}$$

the probability that both will strike is given by the product since the events are independent. Thus

$$\text{Prob (fire fighter's strike and that police's officer strike)} = \frac{1}{21}. \quad \text{(D.1)}$$

(b) What is the probability that neither the police nor the fire fighters will strike?

Prob(fire fighters do not strike and that police officers do not strike)

$$= \left(1 - \frac{1}{3}\right)\left(1 - \frac{1}{7}\right) = \frac{2}{3}\frac{6}{7} = \frac{4}{7}. \quad \text{(D.2)}$$

(c) What is the probability that the police will strike and the fire fighters will not?

Prob(fire fighters do not strike and that police officers strike)

$$= \left(1 - \frac{1}{3}\right)\frac{1}{7} = \frac{2}{3}\frac{1}{7} = \frac{2}{21}. \quad \text{(D.3)}$$

(d) What is the probability that the fire fighters will strike and the police not?

Prob(fire fighters strike and that police officers do not strike)

$$= \frac{1}{3}\left(1 - \frac{1}{7}\right) = \frac{1}{3}\frac{6}{7} = \frac{2}{7}. \tag{D.4}$$

(III) The city water department estimates that, for any day in January, the probability of a water main freezing and breaking is 0.2. For six consecutive days, no water mains break.

(a) What is the probability of this happening? A.

Prob(water main does not break) $= 1 -$ Prob(water breaks)

$$= 1 - 0.2 = 0.8. \tag{D.5}$$

The breaking of a water main in different days are independent events, so also the no break events. Thus

Prob(water main does not break for six days)

$$= \text{Prob(water main does not break for one day)}^6$$

$$= 0.8^6 = 0.262. \tag{D.6}$$

(b) Are the events of breaking independent? A: Yes.

(IV) Jane Watson is running a youth recreation program to reduce the juvenile crime rate. If Jane can get a Law Enforcement Assistance Administration grant to expand the program, she feels that there is a 0.9 probability that the program will work. If she fails to get the grant, the probability of success falls to 0.3. If the probability of getting the grant is 0.6, what is the probability that Jane's program will be successful? A. This problem deals with conditional probabilities. Let us first define them for two general events A and B.

Definition D.4.
Given two events A and B, with Prob$(B) \neq 0$, the probability that A occurs if the event B took place is the *conditional probability* and is denoted by $p(A|B)$ and defined:

$$p(A|B) = \frac{\text{Prob}(A \cap B)}{\text{Prob}(B)}. \tag{D.7}$$

Thus let us define with A the event: *the program works* and B the event: *Jane gets the grant*. Let us call $\overline{B}$ the opposite event: *Jane does not get the grant*. The data of the problem are $p(A|B) = 0.9$ and $p(A|\overline{B}) = 0.3$. The probability that *Jane gets the grant* is $\text{Prob}(B) = 0.6$ and the probability that she does not get the grant is $\text{Prob}(\overline{B}) = 0.4$. Thus the total probability that the program of Jane works can be found by the following calculations:

$$\text{Prob}(A) = \text{Prob}(A \cap B) + \text{Prob}(A \cap \overline{B})$$

$$= p(A|B)\text{Prob}(B) + p(A|\overline{B})\text{Prob}(\overline{B})$$

$$= 0.9 \times 0.6 + 0.3 \times 0.4 = 0.66.$$

(V) Geometric probability. Consider the interval $(0,1)$ and extract a number x from it with uniform probability.

 (a) What is the probability that $x > 1/2$? A. $1/2$
 (b) what is the probability that $1/4 < x < 1/2$? A. $1/4$
 (c) what is the probability that $x = 0$? A. 0.

(VI) Consider a sequence of three numbers x_1, x_2, x_3 taken from the interval $(0,1)$ with uniform probability and suppose that the extractions are done independently.

 (a) What is the probability that $x_1 > 1/2$, $x_2 > 1/2$ and $x_3 > 1/2$? A. $1/8$.
 (b) What is the probability that $x_1 > 1/2$, $x_2 < 1/2$ and $x_3 > 1/2$? A. $1/8$.
 (c) What is the probability that $x_1 > 1/2$, $1/4 < x_2 < 1/2$ and $3/4 < x_3$?
     A.
$$\frac{1}{2}\frac{1}{4}\frac{1}{4} = \frac{1}{32}.$$

(VII) Consider a sequence of numbers $x_1, \ldots, x_N$ taken with uniform probability from the interval $(0,1)$. Define a new random variable y_i: if $x_i > 1/2$ then $y_i = +1$ if $x_i < 1/2$ then $y_i = -1$.

 (a) What is the probability of having a chain of $+1$ of length $n < N$? A: $(1/2)^n$.
 (b) What is the average value $\overline{y}$ of the y_i? A:

$$\overline{y} = \frac{1}{N}\sum_{i=1}^{N} y_i \to 1\text{Prob}(y_i = +1) - 1\text{Prob}(y_i = -1) = 1/2 - 1/2 = 0$$

$$\text{(D.8)}$$

(VIII) Suppose that we have N numbered spheres $m = 1, \ldots, N$ and we extract randomly k of them. We denote with

$$\binom{N}{k}$$

the fraction $\frac{N!}{k!(N-k)!}$ which gives all the possible ways to take k elements from the N elements.

(a) In how many ways is it possible to extract k spheres from the set of N spheres? A.

$$\binom{N}{k}$$

(b) Compute the sum

$$\sum_{k=0}^{N} \binom{N}{k} .$$

A: $2^N = (1+1)^N = \sum_{k=0}^{N} \frac{N!}{k!(N-k)!}$ from the formula of the Newton binomial.

(c) What is the probability $\mathrm{Prob}(k)$ of extracting k spheres? A:

$$\mathrm{Prob}(k) = \frac{\binom{N}{k}}{\sum_{k=0}^{N} \binom{N}{k}}$$

$$= \frac{\binom{N}{k}}{2^N} = 2^{-N} \binom{N}{k} \tag{D.9}$$

(d) Show that $\sum_{k=0}^{N} \mathrm{Prob}(k) = 1$. A: $\sum_{k=0}^{N} \mathrm{Prob}(k) = \sum_{k=0}^{N} 2^{-N} \binom{N}{k}$ but $2^N = \sum_{k=0}^{N} \binom{N}{k}$ and so we get 1.

(e) Compute the mean value of k, $\overline{k}$. A:

$$\overline{k} = \sum_{k=0}^{N} k \mathrm{Prob}(k) = 2^{-N} \sum_{k=0}^{N} k \binom{N}{k} = 2^{-N} \sum_{k=1}^{N} k \binom{N}{k}$$

$$= 2^{-N} \sum_{k=1}^{N} \frac{N!}{(k-1)!(N-k)!}$$

substituting $l = k - 1$ we get

$$= 2^{-N} \sum_{l=0}^{N-1} \frac{N!}{l!(N-l-1)!} = 2^{-N} N \sum_{l=0}^{N-1} \frac{(N-1)!}{l!((N-1)-l)!}$$

$$= \frac{N}{2^N} 2^{N-1} = \frac{N}{2}.$$

(IX) Poisson random variable. Another important meaning of a Poisson random variable is that it can be identified with the number of cells or molecules contained in a given volume V. What is the meaning of the activity λ in this case? A. The Poisson distribution can be written in this case as

$$\text{Prob}(k \text{ cells} \in V) = \frac{(\lambda|V|)^k}{k!} e^{-\lambda|V|}$$

where V is a given region of the space and $|V|$ is the corresponding volume. Let us compute the average number $\bar{k}$ of cells:

$$\bar{k} = \sum_{k=0}^{\infty} k \text{Prob}(k \text{ cells} \in V) =$$

$$= \sum_{k=0}^{\infty} k \frac{(\lambda|V|)^k}{k!} e^{-\lambda|V|}$$

$$= \lambda|V| e^{-\lambda|V|} \sum_{k=1}^{\infty} \frac{(\lambda|V|)^{k-1}}{(k-1)!}$$

$$= \lambda|V| e^{-\lambda|V|} \sum_{k=0}^{\infty} \frac{(\lambda|V|)^k}{k!} = \lambda|V|$$

so we get $\lambda = \bar{k}/|V|$ and the meaning of λ is the density of cells.

(X) Suppose that N spikes arrive to a neuron in the interval $(0, t)$, i.e. $N(t) = N$. Suppose that $N(t)$ is a Poisson variable with activity λt. Suppose also that the spikes have equal probability to be EPSP or IPSP.

(a) Compute the probability of having m EPSPs and $N - m$ IPSPs in the interval $(0, t)$. A

$$\text{Prob}(\#\text{EPSP} = m, \#\text{IPSP} = N - m, \text{in the interval}(0, t))$$

$$= \frac{(\lambda t)^N}{N!} e^{-\lambda t} \left(\frac{1}{2}\right)^m \left(\frac{1}{2}\right)^{N-m} \binom{N}{k} \tag{D.10}$$

This formula can be easily explained: the first factor is the probability of having N spikes in the interval $(0, t)$, the second, $(\frac{1}{2})^m(\frac{1}{2})^{N-m}$, is the probability of having a sequence of m EPSPs and and $N - m$ IPSPS and the last factor $\frac{N!}{m!(N-m)!}$ counts all the possible ways the m EPSPs and $N - m$ IPSPs can distribute in the sequence of N spikes.

(b) Verify the normalization condition

$$\sum_{m=0}^{N} \text{Prob}(\#\text{EPSP} = m, \#\text{IPSP} = N - m, \text{in the interval}(0, t))$$

$$= \frac{(\lambda t)^N}{N!} e^{-\lambda t}. \tag{D.11}$$

A. The sum

$$\sum_{m=0}^{N} \left(\frac{1}{2}\right)^m \left(\frac{1}{2}\right)^{N-m} \binom{N}{k}$$

is the Newton formula applied to the power

$$\left(\frac{1}{2} + \frac{1}{2}\right)^N = 1^N = 1.$$

(XI) The foreign service exam gives a passing grade to only the top 10% of those taking the exam. The mean score is 84, with a standard deviation of 8. What should be the minimum passing grade? A. Let us call the score S. Its mean value $ES = 84$ and its standard deviation $\sqrt{E(S - ES)^2} = 8$. The probability distribution of the variable S is a Gaussian $N(64, 8)$ variable according to Lemma 4.7 of chapter 4. A general property of an $N(\mu, \sigma)$ Gaussian random variable S is that $\text{Prob}(S \in (\mu - 3\sigma, \mu + 3\sigma) \sim 1)$. In our case it should be:

$$\text{Prob}(S \in (84 - 3 \times 8, 84 + 3 \times 8)) = \int_{60}^{108} e^{-(u-84)^2/(2\times 64)} \frac{du}{8\sqrt{2\pi}} \sim 1 \tag{D.12}$$

since $\mu = ES = 84$, $\sigma = \sqrt{E(S - ES)^2} = 8$. This property can be checked numerically with the program which simulates the Gaussian random variables given in this appendix. Thus the maximum score S is 108, since only the top 10% of those passing the exam are passing the minimum passing grade should be

$$\text{minimum passing grade} = 108 - 10\% \times 108 = 97.2 \sim 97. \qquad (D.13)$$

D.2   Program for simulating the $U(0, 1)$ random variables

```
clear all
close all
number_samples=500;
v=zeros(1,number_samples);
x=rand(size(v));
m=mean(x);
errmean=m-0.5;
d=var(x);
errvar=d-1/12;
figure(1);
hist(x)
[f,y]=hist(x);
figure(2);
subplot(2,2,1)
plot(0.1,1.5)
hold on
plot(x)
hold off
subplot(2,2,2)
hist(x)
subplot(2,2,3)
bar(y,f/number_samples);
subplot(2,2,4)
cdfplot(x);
```

We comment only the new commands and functions used in this program.

- 5. **rand** makes a vector x with number_samples components each one being a $U(0, 1)$ r.v.

- 6. the command **mean**(x) makes the average of all the components of the vector x.
- 7. the error made by taking this mean to be the real average of the $U(0, 1)$ is computed, the real average being $1/2$.
- 8. the command **var** computes the variance of the vector x according to the formula given in the main text.
- 9. the same as the point 7. The error **var**$-1/12$ is computed, $1/12$ being the variance of the $U(0, 1)$ random variable.
- 10. the command **figure(1)** gives the name figure 1 to the graph created with the command in the next line.
- 11. the command **hist**(x) creates the graph with the populations in the various sub-intervals of amplitude 0.1 as defined in the main text.
- 12. the command $[f, y]=$**hist**(x) puts the populations created in the vector f.
- 14. the command **subplot(2,2,1)** creates the sequence of four windows that are displayed in Fig. 4.3. This is the one in left upper corner.
- 15. the command **plot**(0.1,1.5) is introduced in such a way that the highest value on the y-axis is 1.5 and the $U(0, 1)$ variables shown in the next plot do not fill all the picture.
- 16. hold on is used for having the next plot on the same plane.
- 17. the command **plot(x)** designs for each integer i the corresponding x_i in the plane.
- 18. **hold off** is necessary for avoiding that the next graphs will be made in the same picture.
- 20. the populations are again plotted for completeness.
- 22. **bar(y,f/number_samples)** plots the histogram of the frequencies i.e. the populations normalized by N.
- 24. the command **cdfplot(x)** designs the empirical distribution function obtained by the sample $(x_1, \ldots, x_N)$ i.e. the fraction

$$\nu_N(x) = \frac{\#\{i|x_i \leq x\}}{N}. \tag{D.14}$$

D.3 Program for simulating the exponentially distributed r.v.

clear all
close all

```
number_samples=1000;
lambda=5;
v=zeros(1,number_samples);
x=rand(size(v));
z=-log(1-x)/lambda;
figure(1);
plot(z)
pause(1);
figure(2);
hist(z)
m=mean(z);
d=var(z);
errm=m − 1/lambda;
errd=d − 1/(lambda)²;
pause(1);
[f, y]=hist(z);
figure(3);
bar(y,f/number_samples);
pause(1);
figure(4);
cdfplot(z);
figure(5);
subplot(2,2,1)
hold on
plot(x)
hold off
subplot(2,2,2)
hist(x)
subplot(2,2,3)
bar(y,f/number_samples);
subplot(2,2,4)
cdfplot(z)
```

The commands are the same as in the previous case the only difference is that the command $z = -\log(1 - x)/\lambda$ is given for generating the exponential random variable x being a $U(0, 1)$ r.v. The error on the mean and variance is given by different formula $m - 1/\lambda$, $d - 1/(\lambda)^2$, where m, d are respectively the empirical mean and dispersion of the sample of the 1000 variables constructed with this program.

D.4 Program for simulating the Gaussian $N(0,1)$ r.v.

We first give the program for simulating the erf(x) function

```
clear all
close all
number_samples=100;
h=1/number_samples;
v=zeros(1,number_samples);
x=rand(size(v));
for i=1:1001
x(i)=(-501+i)*h
v(i)=erf(x(i));
end
figure(1);
plot(0,1.1)
hold on
plot(0, -1.1)
plot(x,v)
```

This program computes the function erf(x) in the interval $(-5, 5)$ and plots the graph of Fig. 4.6. This is the program which generated Fig. 4.7.

```
clear all
close all
number_samples=10000;
v=zeros(1,number_samples);
x=rand(size(v));
v=sqrt(2)*erfinv(2*x-1);
figure(1);
plot(v)
pause(1);
figure(2);
hist(v)
[f,y]=hist(v(1:number_samples));
figure(3);
bar(y,f/number_samples);
pause(1);
m=mean(v);
d=var(v);
```

```
[muhat, sigmahat, muci, sigmaci]=normfit(v);
figure(4);
normplot(v)
figure(5);
subplot(2,3,1)
plot(v)
subplot(2,3,2)
hist(v)
subplot(2,3,3)
bar(y,f/number_samples);
subplot(2,3,4)
cdfplot(v);
subplot(2,3,5)
normplot(v);
```

The **normplot(v)** command produces the test of the hypothesis that the constructed sample is gaussian.

D.5   Program for simulating the Poisson random variables

```
clear all
close all
lambda=2;
number_samples=1000;
for k=1:number_samples
x(k)=rand;
n=0;
sum=0;
sum1=1;
poisson=0;
while poisson==0
if (x(k)*exp(lambda)-sum)*(sum1-x(k)*exp(lambda))>0
poisson=1;
N(k)=n;
end
n=n+1;
sum=sum+lambda^{n-1}/fact(n-1);
sum1=sum+lambda^{n}/fact(n);
```

```
end
end
[f, y]=hist(N(1:number_samples));
z=f/number_samples;
sum1=0;
sum2=0;
for i=1:number_samples
sum1=sum1+N(i);
sum2=sum2+N(i)²;
end
sum1=sum1/number_samples;
sum2=sum2/number_samples;
mean=mean(N);
var=var(N);
err1=lambda-mean;
err2=lambda-var;
err3=(sum2-(sum1)²-sum1)/(sum1)²;
q=length(z);
for k=1:length(z)
p(k)=exp(-lambda)(lambda)^{k-1}/fact(k-1);
w(k)=abs(p(k)-z(k));
end
subplot(2,2,1)
hist(N)
subplot(2,2,2)
bar(y,f/number_samples);
subplot(2,2,3)
plot(p)
subplot(2,2,4) plot(w)
```

- 3. The activity parameter λ of the Poisson r.v. is set equal to 2.
- 4. The number_samples is set equal to 1000 this means that we are going to produce a sample of 1000 Poisson random variables with activity λ.
- 5. Beginning of the loop which constructs these 1000 variables.
- 6. A X $U(0, 1)$ r.v. is chosen.
- 7. n is the possible value of the Poisson variable we start from $n = 0$ and then at the line 14 the current value of n is assigned to the r.v. $N(k)$ if the condition at the line 12 is satisfied.
- 8. The variable sum is initialized. It corresponds to the sum $e^{-\lambda} \sum_{m=0}^{n-1} \frac{\lambda^m}{m!}$

of the lemma. If $n = 0$ this sum is set equal to 0 so that there are no spikes arriving and the variable sum is zero, it means $N(k) = 0$ then at each iteration of the procedure sum is incremented by the term $\frac{\lambda^n}{n!}$.

- 9. The variable sum1 corresponds to the superior term $e^{-\lambda} \sum_{m=0}^{n} \frac{\lambda^m}{m!}$ and is incremented in the same way as the variable sum.

- 10. Poisson is a parameter which defines the fact that the condition of the lemma is satisfied

$$e^{-\lambda} \sum_{m=0}^{n-1} \frac{\lambda^m}{m!} \le X \le e^{-\lambda} \sum_{m=0}^{n} \frac{\lambda^m}{m!}.$$

If this condition is satisfied poisson is set equal to 1 and the loop over n is interrupted otherwise it is equal to zero and the loop over n, initiated by the while condition at line 11, continues.

- 11. **while** poisson==0 initiates the search of the value of n for which the condition $e^{-\lambda} \sum_{m=0}^{n-1} \frac{\lambda^m}{m!} \le X \le e^{-\lambda} \sum_{m=0}^{n} \frac{\lambda^m}{m!}$ holds. The condition is written in the equivalent form $\sum_{m=0}^{n-1} \frac{\lambda^m}{m!} \le e^{\lambda} X \le \sum_{m=0}^{n} \frac{\lambda^m}{m!}$. In this case $X = x(k)$, the k-th variable of the sample, and the condition appears as (x(k)*exp(lambda)-sum)*(sum1-x(k)*exp(lambda))> 0 since sum= $\sum_{m=0}^{n-1} \frac{\lambda^m}{m!}$ and sum1 $= \sum_{m=0}^{n} \frac{\lambda^m}{m!}$.

- 12. **if**(x(k)*exp(lambda)-sum)*(sum1-x(k)*exp(lambda))>0. It is the above discussed condition written in the form of an **if**.

- 13. **poisson**=1 if the condition on $x(k)$ is satisfied the poisson variable is set equal to 1 and so the condition of the **while** poisson==0 is no more true and the loop is interrupted.

- 14. In this case the n appearing in sum and sum1 is the good one and so the value of the Poisson variable is set equal to n and the program goes to the end of the **while** poisson==0 condition.

- 15. The **if** condition acts till this command which must be signed by an **end**.

- 16. If the loops goes on then n is incremented of 1: $n = n + 1$.

- 17. The sum is incremented with the value of the new term with $n = n+1$.

- 18. The sum1 is incremented with the new term.

- 19. **end** is the end of the loop generated by the **while** poisson==0 condition. If the loop has not been interrupted now the program goes to the line 12 where it verifies if with the augmented sum and sum1 the condition (x(k)*exp(lambda)-sum)*(sum1-x(k)*exp(lambda))>0 is verified or not.

- 20. This **end** is the end of the loop over k, after this loop is over the sample $N(k)$ is completed and the various statistics estimates and graphs can be done.

- $20 - 21$. The frequencies of the sample are computed.
- $22 - 32$. The empirical averages of $N(k)$ and $N(k)^2$ are computed.
- $33 - 34$. The mean and var computed on the sample by the matlab functions must coincide with λ if the sample is a Poissonian one.
- 35. If the sample is a good approximation of a Poissonian variable then EN^2 should be equal $(EN)^2 + (EN)$ and so the relative error $(EN^2 - (EN)^2 - EN)/(EN)^2$ is computed.
- 36. The vector z computed at the line 21 has as many components as the number of values of the sample $N(k)$. This number is obtained by this simulation program and so the number of components of z are not known a priori. The instruction **length(z)** gives this number.
- $37 - 40$. If the components of the vector of the frequencies z coincide with those of the Poisson distribution w then the sample constructed with this program is a good approximation of a Poissonian sample. So in this loop the modulus of the difference of the values of the two distributions are computed.
- $41 - 49$. The graph of the populations, the histogram, the one of the true Poisson distribution and the one of the modulus of the differences are done.

Appendix E

Complements to Chapter 5

E.1 Matlab program for simulating the process of Lemma 5.2

```
clear all
close all
numero_spike=100;
lambda=3;
soglia=5;
for i=1:100
v=zeros(1,numero_spike);
tspike=zeros(size(v));
x=rand(size(v));
z=-log(1-x)/lambda;
mp=mean(z);
varp=var(z);
cvt=sqrt(varp)/mp;
t_spike(1)=z(1);
N(1)=0;
for n=1:numero_spike-1
t_spike(n+1)=t_spike(n)+z(n);
N(n+1)=N(n)+1;
end
t_maxtime=max(t_spike);
delta=t_maxtime/1000;
Poisson_Proc(1)=0;
Delta_Poisson_Proc(1)=0;
for n=1:1000
```

```
t=n*delta;
for k=1:numero_spike-1
if (t-t_spike(k))*(t_spike(k+1)-t)>= 0
Poisson_Proc(n)=N(k);
end
end
end
for n=2:1000
Delta_Poisson_Proc(n)=Poisson_Proc(n)-Poisson_Proc(n-1);
end
k=1;
spike=0;
while spike==0&k<=numero_spike
if N(k)>soglia
t_interspike_exc(i)=t_spike(k);
spike=1;
end
k=k+1;
end
end
m_t_interspike_exc=mean(t_interspike_exc);
var_t_interspike_exc=var(t_interspike_exc);
CV_t_interspike_exc=sqrt(var_t_interspike_exc)/m_t_interspike_exc;
err=(1+soglia)/(lambda*m_t_interspike_exc)-1;
err1=sqrt(1+soglia)*CV_t_interspike_exc;
m=0;
index=0;
while index==0
if (m_t_interspike_exc -m)( m+1 -m_t_interspike_exc)>0
int=m;
index=1;
end
m=m+1;
end
fact=fact(m);
for t=1:20
s=t/5;
p(t)=lambda * (lambda * s)^m*exp(-lambda*s)/fact;
```

```
end
subplot(2,4,1)
plot(z)
subplot(2,4,2)
hist(z)
subplot(2,4,3)
plot(t_interspike_exc)
subplot(2,4,4)
hist(t_interspike_exc)
subplot(2,4,5)
stairs(N)
subplot(2,4,6)
stairs(Poisson_Proc)
subplot(2,4,7)
hist(Delta_Poisson_Proc)
subplot(2,4,8) plot(p)
```

E.2   Matlab program for simulating the case of two input Poisson processes

The program in this case is different because for constructing the two Poisson processes a function is used which is displayed after this program.

```
clear all
close all
numero_spike_in=100;
lambda=3;
soglia=5;
tau=2;
for statistica=1:1000
for i=1:2
lambda=i+1;
[Poisson_Proc_out(i,:), Delta_Poisson_Proc_out(i,:), delta_out(i)]=
           =function_poisson(numero_spike_in,lambda,soglia);
end
V(1)=0;
min_delta=min(delta_out);
n=1;
```

```
spike=0;
while spike==0&n<=numero_spike_in*10
V(n)=Poisson_Proc_out(2,n)-Poisson_Proc_out(1,n);
if V(n)>soglia
t_interspike_many_poisson(statistica)=n*min_delta;
spike=1;
end
n=n+1;
end
end
m_t_interspike_many_poisson=mean(t_interspike_many_poisson);
var_t_interspike_many_poisson=var(t_interspike_many_poisson);
err_mean=soglia-m_t_interspike_many_poisson;
err_var=soglia*5-var_t_interspike_many_poisson;
figure hist(t_interspike_many_poisson)

function[Poisson_Proc, Delta_Poisson_Proc, delta] =
                =function_poisson(numero_spike,lambda,soglia)
Poisson_Proc=zeros(1,1000);
Delta_Poisson_Proc=zeros(1,1000);
v=zeros(1,numero_spike);
t_spike=zeros(size(v));
x=rand(size(v));
z=-log(1-x)/lambda;
mp=mean(z);
varp=var(z);
cvt=sqrt(var_p)/mp;
t_spike(1)=z(1);
N(1)=0;
for n=1:numero_spike-1
t_spike(n+1)=t_spike(n)+z(n);
N(n+1)=N(n)+1;
end
t_max_time=max(t_spike);
delta=t_max_time/1000;
Poisson_Proc(1)=0;
Delta_Poisson_Proc(1)=0;
for g=2:1000
t=g*delta;
```

```
for k=1:numero_spike-1
if (t-t_spike(k))*(t_spike(k+1)-t)>=0
Poisson_Proc(g)=N(k);
Delta_Poisson_Proc(g)=Poisson_Proc(g)-Poisson_Proc(g-1);
end
end
end
```

E.3   Matlab program for solving the system (5.22)

```
clear all
close all
numero_spike=100;
lambda=10;
soglia=5;
tau=2;
a=1;
v=zeros(1,numero_spike);
t_spike=zeros(size(v));
for i=1:100
v=zeros(1,numero_spike);
t_spike=zeros(size(v));
x=rand(size(v));
z=-log(1-x)/lambda;
m_p=mean(z);
var_p=var(z);
cvt=sqrt(var_p)/m_p;
t_spike(1)=z(1);
N(1)=0;
for n=1:numero_spike-1
t_spike(n+1)=t_spike(n)+z(n);
N(n+1)=N(n)+1;
end
t_max_time=max(t_spike);
delta=t_max_time/1000;
Poisson_Proc(1)=0;
Delta_Poisson_Proc(1)=0;
for n=1:1000
```

```
t=n*delta;
for k=1:numero_spike-1
if (t-t_spike(k))*(t_spike(k+1)-t)>= 0
Poisson_Proc(n)=N(k);
end
end
end
for n=2:1000
Delta_PP(n)=Poisson_Proc(n)-Poisson_Proc(n-1);
end
V(1)=0;
h=delta/tau;
k=0;
for n=2:1000
t=n*delta;
V(n)=V(n-1)-V(n-1)*h+a*Delta_PP(n);
if V(n)>soglia
k=k+1;
t_interspike_exc(k)=t;
if n<1000
V(n)=0;
end
end
end
subplot(2,4,1)
plot(z)
subplot(2,4,2)
hist(z)
subplot(2,4,3)
plot(V)
subplot(2,4,4)
stairs(Poisson_Proc)
subplot(2,4,5)
stairs(Delta_PP)
subplot(2,4,6)
hist(Delta_PP)
subplot(2,4,7)
plot(t_interspike_exc)
```

```
subplot(2,4,8)
hist(t_interspike_exc)
```

Appendix F

Microarrays

F.1 Measuring gene expression

It is widely believed that thousands of genes and their products, i.e. RNA and proteins, function in a complicated and orchestrated way in a given living organism; therefore technologies which provide a better picture of the interactions among thousands of genes simultaneously are mandatory. Microarray technology allows to miniaturize $mRNA$ abundance measurements onto a single chip so it is possible to study thousands of genes simultaneously. Its tremendous capability for global, parallel gene expression monitoring makes it an ideal starting point for target discovery, enabling a quick assessment of what differentiates the transcriptional profiles of different tissues or cell types.

Microarrays are based on the complementary binding (hybridization) of the unknown sequences. With microarrays, thousands of spotted samples known as probes (with known identity) are immobilized on a solid support (a microscope glass slides or silicon chips or nylon membrane). When fluorescently-labeled cDNA (from an mRNA template) hybridizes to the DNA probe on the solid surface, the location of the DNA probe on the microarray surface emits a fluorescent signal. The intensity of the signal on the fluorescent array is relatively proportionate to the amount of fluorescently-labeled cDNA bound to the probe.

The lengths of the probes on the arrays vary from manufacturer to manufacturer and probe lengths ranges from 25 to 50 nucleotides, or 80 nucleotides. The length of the probe, as well as the sequence selected for each probe, greatly affects the probe performance (i.e. its ability to uniquely detect a specific transcript).

At this moment four basic types of microarrays exist:

(1) Spotted arrays or cDNA-microarrays are small glass slides on which pre-synthesized single stranded DNA or double-stranded DNA is spotted. These DNA fragments are usually several hundred base pairs in length and are derived from ESTs (Expressed Sequence Tag) or known transcribed sequences from organism studied. Usually each spot represents one single ORF (Open Reading Frame) or gene. A pair of cDNA samples is independently copied from the corresponding *mRNA* populations (usually derived from a reference and a test sample) with reverse transcriptase and labeled using distinct fluorochromes (green and red). These cDNA samples are subsequently pooled and hybridised to the array. Relative amounts of a particular gene transcript in the two samples are determined by measuring the signal intensities detected for both fluorochromes and calculating the ratios (here, only relative expression levels are usually obtained). A cDNA microarray is therefore a differential technique, which intrinsically normalizes for noise and background.

(2) GeneChip oligonucleotide arrays (Affymetrix) are high-density arrays of oligonucleotides synthesized in situ using light-directed chemistry consisting of thousands of different oligomer probes (25-mers). Each gene is represented by $3 - 20$ different oligonucleotides, serving as unique sequence-specific detectors. In addition mismatch control oligonucleotides (identical to the perfect match probes except for a single base-pair mismatch) are added. These control probes allow estimation of cross-hybridization. With this technology, absolute expression levels are obtained (no ratios).

(3) Applied Biosystem arrays are single channel oligo arrays developed for human, mouse and rat genomes. Probes are 60mers oligonucleotides for the optimum combination of sensitivity and specificity. This permits the single-base hybridization specificity of shorter oligonucleotides without the decrease sensitivity encountered when using shorter probes. Probes are located within the last 1500 bp of the transcript end. Whenever possible, probes are designed against common regions of multiple alternative transcripts for a given gene. When alternative transcripts have different ends additional probe designs are selected.

The data set includes genes from both the public database and the Celera Genomics database.

(4) Illumina arrays are based on the use of oligonucleotides (50mers) synthetized on beads. Illumina employs two different types of Arrays

platforms: the microplate-compatible Sentrix Array Matrix, with 96 fiber optic array bundles and the Micro-Electro-Mechanical Systems (MEMS)-patterned Sentrix BeadChip, which can be etched in many different configurations.

The manufacture of the arrays consists of three parts. The first part is the creation of a master bead pool consisting of 1536250000 different bead types. Oligonucleotide capture probes are immobilized individually by bead type in a bulk process. Each bead type in an array comes from a single immobilization event, reducing array-to-array feature variability. The second step is the random self-assembly of the master pool of bead types into etched wells on the array substrate, where each bead type has an average $30\times$ representation (a strategy that provides the statistical accuracy of multiple measurements). The third step is the identification of each bead on the array by means of a decoding process. A combinatorial series of brief hybridizations and rinses is used that result in a level of accuracy well beyond the requirements of any application. This decoding process also provides a quality control measure of the function of each bead that is incorporated into the final bead map.

After microarray hybridization experiments, laser scanning of slide produces an image from which the raw intensity data has to be extracted. A variety of software tools have been developed for image analysis. The basic idea is to get the varying intensities for each spot in an array. Although this is a relatively straightforward goal, there is as yet no common manner of extracting this information, and many research groups are still writing customized software for this purpose. Scanning and image processing are currently resource-intensive tasks, requiring human intervention to ensure that grids are properly aligned and that artifacts are flagged and properly excluded from subsequent analysis.

Image analysis involves three stages. First, the arrayed genes must be identified from spurious signals that can arise due to precipitated probe or other hybridization artifacts or dust on the surface of the slide. After gridding the spot intensities (real signal), background (noise) has to be calculated for spots. It is always better to calculate background locally for each spot, rather than globally for the entire image. Next step in image processing is the extraction of signal, noise and quality control measures for the spot.

The expression data from a microarray experiment are very noisy in nature, in each step of the preparation of the test and reference sample and at each stage in the process errors could be introduced. Therefore it is necessary to apply transformation to express data in such a way to balance the individual hybridization intensities appropriately so that meaningful biological comparisons can be made. This process is called normalization. There are a number of reasons why data must be normalized, including unequal quantities of starting RNA, differences in labeling or detection efficiencies between the fluorescent dyes used, and systematic biases in the measured expression levels. Hence the aim of normalization is to remove the effect of this noise from the data, while still maintaining the ability to detect significantly differentially expressed genes. The commonly used normalization procedures are: total intensity normalization, LOWESS normalization, mean centering, ratio statistics and standard deviation regularization.

The next step is to organize, analyze and visualize this data to reach conclusions about the biological processes being studied. The methods which can be used to achieve this are as open ended as the range of biological processes available to study. To date, a variety of methods have been used including:

- clustering by correlation/mutual information [Eisen (1998)],[Spellman (1998)]
- graph-based/hierarchical clustering [Ben-Dor (1999)]
- Gaussian mixture model clustering [Yeung (2003)]
- self-organizing maps
- dimensionality reduction (PCA, ICA) [Raychaudhuri (2000)]

F.2   Applications of microarray

The range of applications of microarray technology is enormous. A powerful application of microarray technology is tumor types classification and a number of studies have successfully utilized this approach in human cancer. We can also construct a temporal expression profile for the thousands genes present on the array by measuring the evolution of mRNA levels at consecutive time points during a biological experiment.

The list of potential uses of this technique is not limited to cancer research. For example, the temporal impact on gene expression by drugs, environmental toxins, or oncogenes may be elucidated, and regulatory networks and co-expression patterns can then be deciphered.

A powerful means of analyzing microarray expression data is the identification of groups of genes that are expressed in similar patterns. This technique is valuable because co-expressed genes might be controlled by the same regulatory mechanism. Several computational methods have been developed to identify patterns of co-expression. These include hierarchical clustering, Bayesian clustering, K-mean clustering, Pearson rank clustering, Self-Organizing Maps (*SOMs*), Support Vector Machines (*SVM*).

A correlated approach of the analysis of microarray expression data is the identification of genes which are expressed differently (for example: up- or down-regulated) in response to experimental conditions. Because genes that are highly regulated in response to a particular stimulus are likely to have important roles in the corresponding cellular response, this approach can be useful in correlating functions with genes. However, one limitation of this method is that it requires an a priori determination of a threshold level of significance for expression ratios. If this threshold is set too high, some relevant genes will be ignored. Conversely, a threshold that is set too low will lead to false correlations and the misclassification of genes. In practice, changes in expression of at least 1.0 fold have been considered significant, although only rarely has any statistical or biological justification been offered for the selection of this threshold.

Appendix G

Complements to Chapter 6

G.1 Kohonen algorithm in Matlab source

We describe here the Matlab programs which perform the Kohonen algorithm related to the genetic classification described in chapter 5. Below we have 3 functions:

(1) the first function, called *kohonen_all*, creates a large data set to be presented to the network starting from the original data set and then calls the function *kohonen*.
(2) the second function, called *kohonen*, performs the Kohonen algorithm by exploiting the function *update*.
(3) the third function, called *update*, runs the update rule of the Kohonen algorithm (see 6.12).

The functions have been implemented to cluster the data set illustrated in chapter 5. Therefore they describe how to divide in clusters different data sets of about 2000 elements; these data sets represent the data of different biological samples. We have implemented the kohonen algorithm choosing Λ as the neighborhood function of the *winner* neuron (see 6.15).

To run the Kohonen algorithm it is enough to key in Matlab window the following instruction:

$$[kohdat, weights] = kohonen\_all(dat, ndat, cost, n, dist)$$

where *dat* will be the data set to be clustered, *ndat* the number of elements contained in the data set, *cost* the number of times we want to repeat the elements, *n* the number of weights, *dist* the value of parameter s of Λ which determines the activation region around the *winner* neuron (that is which neurons must update their weight vectors together with the winner neuron).

The outputs of this function are the weights vector which are in the variable *weights*; in *kohdat* there are the weights and the different initial conditions in order to check that actually the algorithm converges independently from these (the initial conditions).

Let us make some comments about the routines.

In the first function from lines 3 to 9 there are the instructions to create a large data set from the original one. Firstly we have permuted randomly the data, subsequently we have put together the original data set with the permuted one. We have repeated this task as many times as the variable *cost* indicates. The other lines call the function *kohonen*.

The function *kohonen* generates randomly the weights at the first iteration of the algorithm ($n = 1$), then calls the function *update*. This task is repeated as many times as the variable *cost_v* indicates in order to check that we always obtain the same weights even if we change the initial values of them.

The function *update* represents the core of the Kohonen algorithm. Here the winner neuron is chosen, that is the neuron with the weight which is far from the data input less all the other weights. Then the weights are updated according to (6.12).

```
function[kohdat,weights]=kohonen_all(dat,ndat,cost,n,dist,nsamples)
dat_t=dat;
for j = 1 : cost
   ind = randperm(ndat);
   for q = 1 : ndat
    dat_random(q,:)=dati(ind(q),:);
   end
   dat_t=cat(1,dat_t,dat_random);
end % for j
ndat_t=ndat*(cost+1);
for i = 1: nsamples
   [kohdati(:,i),weights(i)]=kohonen(dat_t(:,i),ndat_t,n,dist);
end
```

```
function [kohdat,weights]=kohonen(dat,ndat,n,dist)
cost_v= 3;
weights=struct('value',[ ]);
weights_interm=struct('value',[ ]);
```

```
kohdat=struct('val_init_random',[ ],'weights',[ ]);
flag= 1;
for t = 1 : cost_v
  for j = 1 : ndat
    count= j;
      % weights (step 0) %
    if(count == 1)
       count = 2;
       if(flag == 1)
           maximum=ceil(max(dat));
       end
       rand('state',sum(100*clock))
       R=rand(n,1);
       w=maximum*(R/(max(R)+1)) ;
       kohdat(t).val_init_random=w;
       w1=w;
    end %if count
  eta_i= 0.8;
  eta=eta_i * (1- count/ndati);
  w=update(n,eta,dist,w,dat(j),count);
  end % for j
flag= 2;
kohdat(t).weights=w;
end % for t
weights.value= kohdat(t).weights;



function w=update(n_centers,eta,dist,w,z,count)
% searching for winner neuron%
min=200000;
for i = 1 : n_centers
  norm=abs(w(i)-z);
  if (norm<min )
  min=norma;
  winner=i ;
  end
end % for i
% weights update %
for j = 1 : n_centers
```

```
s=abs(j-winner);
 if(s ≤ dist)
   w(j)=w(j) + (eta*(z - w(j))) ;
 end
end %for j
```

Appendix H

Mathematical description of Kohonen algorithms

H.1 Convergence of Kohonen algorithm

The first result about the algorithm convergence was found by Kohonen [Kohonen (1982)]. He concentrated on one-dimensional mapping and demonstrated that the weights converge in mean to the limit values. Although the result is enunciated as a.e. convergence in the article only the convergence in mean is proven. The convergence in mean is obtained by making the average of the weights on many different sequences of patterns $x(n)$. The ordering of the weights has been proved in [Kohonen (1982)] for the winner-take-all process.

In the paper of Erwin et al. [Erwin (1992)], [Erwin (1992)] there is a proof of ordering for one-dimensional case which holds for any neighborhood function which is monotonically decreasing with distance and in the case of non uniformly distributed input.

Many other authors [Lo (1991)], [Lin (1998)], [Ritter (1986)], [Ritter (1988)], [Sadeghi (2001)], [Tirozzi (1997)], [Yin (1995)], investigated the convergence properties of the Kohonen algorithm in one and more dimensions, someone by viewing the weight values as states of a Markov process, others using the ordinary differential equations for the mean values of the network. But the main results have been limited to one-dimensional map where the property of order is valid and under certain conditions on $\eta(n)$, the expectation of the values weights converges to a unique value. The existence and uniqueness of the minimum is ensured by the existence of a unique minimum of some functional, but the existence of the minimum is difficult to check for non-uniform distribution of the input values especially in the multi-dimensional case.

In more than one dimension, despite the robustness of the algorithm which has been used successfully in many different application area, there is still no proof of a necessary and sufficient condition for the convergence of the algorithm. There are proofs of sufficient conditions and only a few for the multi-dimensional case, see for example Feng and Tirozzi [Tirozzi (1997)], Lin and Si [Lin (1998)], Sadeghi [Sadeghi (2001)]. Lin and Si have shown that the distribution of the weight values converges to a stationary state introducing and studying the same objective function proposed by Ritter and Schulten [Ritter (1986)]. In the paper of Feng and Tirozzi the convergence problem of the Kohonen feature mapping algorithm has been proven by using stochastic approximation theory. But in all these papers the rate of decrease of the learning parameter is too fast and so these theorems are contradicted by numerical results. Only in the paper of Feng and Tirozzi it is mentioned explicitly that the rate of decrease of the learning parameter of these theorems is too fast and there is a proposal for a slower decay.

In this appendix we show that there is a.e. convergence if the rate is the one of numerical simulations. In Section 3 we will give many examples of *good* and *bad* decay of $\eta(n)$. The choice of $\eta(n)$ is important also for the speed of convergence of the process. Another key role for the a.e. convergence is the form of the probability distribution of the data as it will be clear from the theorem we present below.

In order to understand it we need some definitions. Let us introduce a function g which is the leading term of the super martingale difference:

$$g(y_1, y_2, ..., y_N; \omega_1, \omega_2, ..., \omega_N) = \sum_{i=1}^{N}(y_i - \omega_i) \cdot \sum_k \left(\int_{\Pi(y)_k} \Lambda(k,i)(x - y_i)f(x)dx \right),$$
(H.1)

where $\omega_i(n) = (\omega_{ij}(n), j = 1, ..., M, i = 1, ..., N)$ and $y_i(n) = (y_{ij}(n), j = 1, ...M, i = 1, ..., N) \in \mathbb{R}^{M \times N}$ f is the density of the probability distribution of the data with support on a compact set Ω of $\mathbb{R}^M$, $\Pi(y)$ is the Voronoi tessellation associated with y (see (6.14)). $(y_i - \omega_i) \cdot (x - y_i)$ is the M-dimensional scalar product.

We define also:

$$\Theta \equiv \{the\ set\ of\ all\ Voronoi\ tessellations\ associated\ with\ \{\omega_1(n), ..., \omega_N(n)\}\}$$

for all n.

For $y \in \mathbb{R}^M$ we use the convention that $y \in \Theta$ implies that there exists a Voronoi tessellation $\Pi(y)$ such that $\{\Pi(y)_i, i = 1, ..., N\} \in \Theta$. Finally we can enunciate our theorem:

Theorem H.1. *Let the vectors* $\omega(n) \in \mathbb{R}^{M \times N}$ *be updated by the Kononen algorithm (6.13)*

$$\omega_i(n+1) = \omega_i(n) + \eta(n)\Lambda(i,v)\bar{\mathbb{I}}(\omega_v(n), \xi(n+1)) \cdot (\xi(n+1) - \omega_i(n))$$

if there exists a unique point $\tilde{\omega} = (\tilde{\omega}_1, ..., \tilde{\omega}_N) \in \mathbb{R}^{M \times N}$ *such that for each* $y = (y_1, y_2, ..., y_N)$:

$$g(y_1, y_2, ..., y_N; \tilde{\omega}_1, ..., \tilde{\omega}_N) \leq 0 \; \forall y \in \Theta \tag{H.2}$$

where the equality holds if and only if $y_i = \tilde{\omega}_i$ $i = 1, ..., N$ *and* :

$$\sum_{n=1}^{+\infty} \eta(n) = +\infty \qquad \lim_{n \to +\infty} \eta(n) = 0 \tag{H.3}$$

then we almost everywhere have:

$$\lim_{n \to +\infty} \omega_i(n) = \tilde{\omega}_i \qquad i = 1, ..., N.$$

Remark H.1. This theorem is interesting because the rate of decay of $\eta(n)$ is the one used in simulations but it is still not enough because the full proposition should exclude the decays which are not used in the simulations i.e. the ones such that

$$\sum_{n=1}^{+\infty} \eta(n)^2 < +\infty.$$

This last condition is often required in the proofs of theorem about the convergence of Kohonen algorithm, but we have checked in our simulation that there is no convergence. For example if we use $\eta(n) = \frac{1}{n}$ the limit values of weights are not ordered at the end of the learning process for any initial condition (that is for any random choice of weights at the beginning of the algorithm). This result contradicts the one of Sadeghi [Sadeghi (2001)]. In his paper he made a numerical check but it is not enough since he has proven directly only the convergence in mean and not the a.e. convergence and in addition in his simulation he started from ordered weights.

Remark H.2. Although the theorem is formulated in the multi-dimensional case we use it in one dimension because the condition (H.2) is not easy to check in the general case. For $M = 1$ it has been seen in the paper [Tirozzi (1997)] that, if the distribution of the data is uniform and the data belong to the interval $(0, 1)$, the clusters are intervals of amplitude 0.1 for $N = 10$. They are centered around the points $(0.5, 1.5, ...)$. If the data are gaussian distributed, as in the biological case, there is no unique point satisfying condition (H.2) and other arguments must be used. We show in Section 3 that, choosing $\eta(n)$ in a particular way, it is still possible to have a.e. convergence but there is no theorem justifying this result.

Bibliography

Altman, R.B. and Dugan, J.M. (2003) *Defining bioinformatics and structural bioinformatics*, in *Structural bioinformatics*, ed. by P.E. Bourne and H. Weissig (2003), pp. 3–14. (John Wiley & Sons).

Altschul, S.F. (1991). *Amino acid substitution matrices from an information theoretic perspective* J. Mol. Biol., Vol. **219**, pp. 555–565.

Altschul, S.F. (1993). *A protein alignment scoring system sensitive at all evolutionary distances* J. Mol. Evol., Vol. **36**, pp. 290–300.

Altschul, S.F., Gish, W., Miller, W., Myers, E.W., Lipman, D.J. (1990). *Basic local alignment search tool*, J. Mol. Biol., Vol. **215**, pp. 403–410.

Anderberg, M.R. (1973). *Cluster Analysis for Applications* (Accademic Press).

Baldi, P., Brunak, S., (1998). *Bioinformatics: The Machine Learning Approach* (MIT Press).

Ben-Dor, Shamir R. and Yakhini Z. (1999). *Clustering Gene Expression Patterns, Proceedings of the Third Annual International Conference on Computational Molecular Biology (RECOMB'99)*, Lyon, France.

Berman, H.M., Westbrook, J., Feng, Z., Gilliland, G., Bhat, T.N., Weissig, H., Shindyalov, I.N., Bourne, P.E. (2000). *The Protein Data Bank*, Nucl. Acids Res., Vol. **28**, pp. 235–242.

Bishop, C.M. (1995).. *Neural networks for Pattern Recognition.* (Oxford University Press).

Bourne, P.H. and Weissig, H. (2003), *Structural bioinformatics*, eds. P.H. Bourne and H. Weissig, John Wiley & Sons, Hoboken, NJ.

Bouton, C. and Pages, G. (1993). *Self-Organization and a.s. Convergence of the One-Dimensional Kohonen Algorithm with Non Uniformly Distributed Stimuli, Stochastic Process Appl*, Vol. **47**, pp. 249–274.

Brannetti, B., Via, A., Cestra, G., Cesareni, G., and Helmer-citterich, M., (2000). *SH3-SPOT: An algorithm to predict preferred ligands to different members of the SH3 gene family*, J. Mol. Biol., Vol. **298**, pp. 313–328.

Brannetti, B. and Helmer-Citterich, M. (2003). *SPOT: A web tool to infer the interaction specificity of families of protein modules*, Nucl. Acids Res., Vol. **31**, pp. 3709–3711.

Carpenter, G., A., and Grossberg, S. (1988). *The ART adaptive pattern recogni-*

tion by a self-organizing neural network, Computer, Vol. **21**, pp. 77–88.

Chauvin, Y. and Rumelhart, D.E. (1995). *Backpropagation: Theory, Architecture and Applications*, eds. Y. Chauvin and D.E. Rumelhart. Lawrence Erlbaum Associates, Hillsdale, NJ.

Collins, S., R., Kemmeren, P., Zhao, X., C., Greenblatt, J., F., Spencer, F., Holstege, F., C., Weissman, J., S., Krogan, N., J., (2007). *Towards a comprehensive atlas of the physical interactome of Saccharomyces cerevisiae*, Mol. Cell. Proteomics, in press. E-pub on 2 Jan 2007.

Cottrell, M. and Fort, J.C. (1987). *Etude d'un processus d'auto-organisation*, Annales de l'Institut Henri Poincar, Vol. **23**, pp. 1–20.

Crooks, G.E., Hon, G., Chandonia, J.M., Brenner, S.E. (2004). *Weblogo: a sequence logo generator*, Genome Res., Vol. **14**, pp. 1188–1190.

Dayhoff, M. O., Schwartz, R. M., and Orcutt, B. C. (1978). *A model for evolutionary change in proteins*, in *Atlas of Protein Sequence and Structure*(Dayhoff M.O. editor), Vol. **5**, pp. 345–352.

Durbin, R., Eddy, S., Krogh, A., Mitchison, G. (1998). *Biological sequence analysis*. (Cambridge University Press).

Eidhammer, I., Jonassen, I., Taylor, W.R. (2004). *Protein Bioinformatics: an algorithmic approach to sequence and structure analysis*. (John Wiley & Sons).

Eisen, M., Spellman, P.T, Brown, P.O. and Botstein D. (1998). *Cluster Analysis and Display of Genome-Wide Expression patterns*, Proc. Natl. Acad. Sci. USA, Vol. **95**, pp. 14863–14868.

Erwin, Ed., Obermayer, K. and Schulten, K. (1992). *Self-Organizing Maps: Stationary States, Metastability and Convergence Rate*, Biological Cybernetics, Vol. **67**, pp. 35–45.

Erwin, Ed., Obermayer, K. and Schulten, K. (1992). *Self-Organizing Maps: Ordering, Convergence Properties and Energy Function*, Biological Cybernetics, Vol. **67**, pp. 47–55.

Ewens, W.J. and Grant, G.R. (2001). *Statistical Methods in Bioinformatics. An Introduction.*, (Springer-Verlag, NY).

Feller, W. (1971). *Probability Theory and its Applications*, Vol. **2** (John Wiley & Sons, second ed.).

Feng, J.F. (2004). *Computational Neuroscience, A Comprehensive Approach*, (Chapman & Hall/CRC).

Feng, J.F. and Tirozzi B. (1997). *Convergence Theorem for the Kohonen Feature Mapping Algorithms with VLRPs*, Computer Math. Applic., Vol. **33**, No.3, pp. 45–63.

Fitzhugh R. (1960). *Thresholds and plateaus in the Hodgkin-Huxley nerve equations*, J. Gen. Phys, Vol. **43**, pp. 867–896.

Fort, J.C. and Pages, G. (1995). *On the a.s. Convergence of the Kohonen Algorithm with a General Neighborhood Function*, The Annals of Applied Probability, Vol. **5**, pp. 1177–1216.

Gattiker, A., Gasteiger, E., Bairoch, A., (2002). *ScanProsite: a reference implementation of a PROSITE scanning tool*, Applied Bioinformatics, Vol. **1**, pp. 107–108.

Gavin, A., C., et al., (2006). *Proteome survey reveals modularity of the yeast cell machinery, Nature*, Vol. **440**, pp. 631–636.

Goldman D.E. (1943). *Potential, impedance and rectification in membranes , J. Gen. Phys*, Vol. **272**, pp. 37–60.

Gondran, M. and Minoux, M. (1979). *Graphes et Algorithmes* (Eyrolles).

Hassard D.D., Kazarinoff N.D., and Wan Y-H. (1981) *Theory and Applications of Hopf Bifurcation* (Cambridge University Press, Cambridge).

Henikoff S., Henikoff J.G. (1992). *Amino acid substitution matrices from protein blocks, Proc. Natl. Acad. Sci.*, Vol. **89**, pp. 10915–10919.

Hertz, J., Krogh, A. and Palmer R. (1991). *Introduction to the Theory of Neural Computation, Lectures Notes of Santa Fe Institute*, (Addison Wesley).

Higgins, D., Thompson, J., Gibson, T., Thompson, J.D., Higgins, D.G., Gibson, T.J.(1994). *CLUSTAL W: improving the sensitivity of progressive multiple sequence alignment through sequence weighting, position-specific gap penalties and weight matrix choice Nucl. Acids Res.*, Vol. **22**, pp. 4673–4680.

Hodgkin A.L. and Katz B. (1949). *The effect of sodium ions on electrical activity of the giant axon of the squid, J. Physiol*, Vol. **108**, pp. 37–77.

Hodgkin A. L., and Huxley (1952a). *Current carried by sodium and potassium ions through the membrane of the giant axon of Loligo , J. Physiol.*, Vol. **116**, pp. 449-472.

Hodgkin A. L., and Huxley (1952b). *The components of membrane conductance in the gian axon of Loligo, J. Physiol.*, Vol. **116**, pp. 473-496.

Hodgkin A. L., and Huxley (1952c). *The dual effect of membrane potential on sodium conductance in the gian axon of Loligo, J. Physiol.*, Vol. **116**, pp. 497-506.

Hodgkin A. L., and Huxley (1952d). *A quantitative description of membrane current and its application to conduction and excitation nerve , J. Physiol.*, Vol. **117**, pp. 500-544.

Irizarry R.A, Bolstad B.M., Collin F., Cope L.M., Hobbs B. and Speed T.P. (2003). *Summaries of Affymetrix GeneChip Probe Level Data, Nucleic Acids Res.*, Vol. **31**.

Krogan, N., J., et al. (2006). *Global landscape of protein complexes in the yeast Saccharomyces cerevisiae, Nature*, vol. 440, pp. 637-643.

Le Cun, Y., (1998). *A theoretical framework for backpropagation, In Proceedings of the 1988 Connectionist Models Summer School.* eds Touretzky, D., Hinton, G., Sejnowski, T., pp. 18–21. Morgan Kauffman, S.Mateo, CA.

Lin, S. and Si, J. (1998). *Weigth-Value Convergence of the SOM Algorithm for Discrete Input, Neural Computation*, Vol. **10**, pp. 807–814.

Lo Z-P. and Bavarian B. (1991). *On the Rate of Convergence in Topology Preserving Neural Networks, Biological Cybernetics*, Vol. **65**, pp. 55–63.

Kohonen, T. (1989). *Self-Organization and Associative Memory Process*, (Springer-Verlag).

Kohonen, T. (1991). *Analysis of a Simple Self-Organizing Process, Biological Cybernetics*, Vol. **44**, pp. 135–140.

Kohonen, T. (1991). *Self-Organizing maps: optimization approaches, Artificial Neural Networks*, Vol. **1**, pp. 891–990.

Marchetti, G. (2006). *Clustering Algorithms and Genetic Data, Thesis at University La Sapienza, Rome.*

McGinnis, S. and Madden, T.L. (2004). *BLAST: at the core of a powerful and diverse set of sequence analysis tools, Nucl. Acid Res.*, Vol. **32** (Web server issue), pp. W20-W25.

Mezard, M., Parisi, G. and Virasoro, M.A. (1987). *Spin Glass Theory and Beyond*, (World Scientific).

Nevel'son M.B. and Has'minskii R.Z. (1976). *Stochastic Approximation and Recursive Estimation, Translation of Math. Monograph 47*, (Amer. Math. Soc).

Olsen, J., V., Blagoev, B., Gnad, F., Macek, B., Kumar, C., Mortensen, P., Mann, M., (2006).*Global, in vivo, and site-specific phosphorylation dynamics in signaling networks, Cell*, Vol. **127**(3), pp. 635–648.

Pantanetti, V. (2006). *Clustering Methods Applied to Genetic Classification , Thesis at University La Sapienza, Rome.*

Pearson, W.R. and Lipman, D.J. (1989). *Improved Tools for Biological Sequence Comparison, Proc. Natl. Acad. Sci.*, Vol. **85**, pp. 2444–2448.

Quaglino, E. and Calogero R. (2004). *Concordat Morphologic and Gene Expression Data Show that a Vaccine Halts HER-2/neu Prenoplastic Lesions, The Journal of Clinical Investigation*, Vol. **113**, No. 5.

Raychaudhuri, S., Stuart, J.M. and Altman, R.B. (2000). *Principal Components Analysis to Summarize Microarray Experiments: Application to Sporulation Time Series, Pac Symp Biocomput*, pp. 455–466.

Ritter, H. and Shulten, K. (1986). *On the Stationary States of Kohonen's Self-Organizing Sensory Mapping, Biological Cybernetics*, Vol. **54**, pp. 99–106.

Ritter, H. and Shulten, K. (1988). *Kohonen's Self-Organizing Maps: Exploring Their Computational Capabilities, Proceedings of the ICNN'88, IEEE International Conference on Neural Networks*, Vol. **1**, pp. 109–116.

Rosenblatt, F. (1962). *Principles of Neurodynamics*, (Spartan Books, Washington, DC).

Rost, B., and Sander, C. (1993). *Improved prediction of protein secondary structure by use of sequence profiles and neural networks, Proc. Natl. Acad. Sci. USA*, Vol. **90**, 7558–7562.

Sadeghi A.A. (2001). *Convergence in Distribution of the Multidimensional Kohonen Algorithm, Journ. of Appl. Prob.*, Vol. **38**, pp. 136–151.

Saviozzi, S., Iazzetti, G., Caserta, E., Guffanti, A. and Calogero, R. (2003). *Microarrays Data Analysis and Mining, Methods in molecular medicine*, Vol. **94**, pp. 67–90.

Schneider, T.D. and Stephens, R.M. (1990). *Sequence logos: a new way to display consensus sequences, Nucl. Acid Res.*, Vol. **18**, pp. 6097–6100.

Schneider, T.D. and Wrede, P. (1994). *The rational design of amino acid sequences by artificial neural networks and simulated molecular evolution: de novo design of an idealized leader peptidase cleavage site, Biophys. Journal*, Vol. **66**, pp. 335-344.

Schutkowski, M., Reineke, U., Reimer, U., (2005). *Peptide arrays for kinase profiling, Chembiochem.*, Vol. **6**(3), pp. 512–521.

Scott, A. C., (1975). *The electrophysics of a nerve fiber, Review of Modern Physics*, Vol. **47**, pp. 487–533.

Shcherbina, M. and Tirozzi, B. (1993). *The Free Energy of a Class of Hopfield Models, J. of Stat. Phys.*, Vol. **72** 1/2, pp. 113–125.

Shcherbina, M. and Tirozzi, B. (2003). *Rigorous Solution of the Gardner Problem, Commun. Math. Phys.*, Vol. **234**, pp. 383–422.

Simon, B. (1979). *Functional integration and quantum physics*, (Academic Press).

Smith, G., P., (1985). *Filamentous fusion phage: novel expression vectors that display cloned antigens on the virion surface, Science*, Vol. **228**, pp 1315–1317.

Smith, G., R., and Sternberg M., J., (2002). *Prediction of protein-protein interactions by docking methods, Curr. Op. Struct. Biol.*, Vol. **12**, pp. 28–35.

Spellman, P.T., Sherlock, G., Zhang, M.Q., Iyer, V.R., Anders, K., Eisen, M.B., Brown, P.B., Botstein D. and Futcher B. (1998). *Comprehensive Identification of Cell Cycle-regulated Genes of the Yeast Saccharomyces Cerevisiae by Microarray Hybridization, Molecular Biology of the Cell*, Vol. **9**,

Steen, H. and Mann, M. (2004). *The ABC's (and XYZ's) of peptide sequencing., Nature Rev. Mol. Cell. Biol.*, Vol. **5**, pp. 699–711.

Tamayo, P., Slonim, D., Mesirov, J., Zhu, Q., Kitareewan, S., Dmitrovsky, E. and Lander, E.S. (1999). *Interpreting Patterns of Gene Expression with Self-Organizing Maps: Methods and application to hematopoietic differentiation, Proc. Natl. Acad. Sci.USA*, Vol. **96**, pp. 2907–2912.

Taylor, J.G. and Budinich, M.(1995). *On the Ordering Conditions for Self-Organising Maps , Neural Computation*, Vol. **7**, pp. 284–289.

Tolat, V.V. (1990). *An Analysis of Kohonen's Self-Organizing Maps Using a System of Energy Functions, Biological Cybernetics* , Vol. **64**, pp. 155–164.

Tuckwell, H.C. (1988). *Introduction to Theoretical Neurobiology*, Vol. **1** (Cambridge University Press).

Tusher V.G., Tibshirani R. and Chu G. (2001). *Significance Analysis of Microarrays Applied to the Ionising Radiotion Response, Proc. Natl. Acad. Sci. USA*, Vol. **98**, pp. 5116–5121.

Valle, G., Helmer-Citterich, M., Attimonelli, M. and Pesole, G. (2003). *Introduzione alla Bioinformatica*, (Zanichelli ed. S.p.A.).

Vesanto, J. and Alhoniemi, E. (2000). *Clustering of the Self-Organizing Map, IEEE Transactions on Neural Networks*, Vol. **11**, pp. 586–600.

Wasinger, V., C., Cordwell S., J., Cerpa-Poljak, A., Yan, J., X., Gooley, A., A., Wilkins, M., R., Duncan, M., W., Harris, R., Williams, K., R., Humphery-Smith, I., (1995). *Progress with gene-product mapping of the Mollicutes: Mycoplasma genitalium, Electrophoresis*, Vol. **16**(7), pp. 1090–1094.

Wu, C.H. (1996). *Artificial neural networks for molecular sequence analysis, Computers Chem.*, Vol. **21**, pp. 237–256.

Yeung, K.Y., Medvedovic M. and Bumgarner R.E. (2003). *Clustering Gene-Expression Data with Repeated Measurements, Genome Biology*.

Yin, H. and Allison N.M (1995). *On the distribution and convergence of feature space in self-organizing maps, Neural Computation*, Vol.**7**, pp. 1178–1187.

Subject Index

Author Index

Ampere, 5
Anderberg, 102

Ben-Dor, 109

Coulomb, 4

Feng, 119
Fitzhugh, vii, 32

Hodgkin, vii
Hopf, 39
Huxley, vii

Joule, 4

Kirchhoff, 14
Kohonen, 110, 113, 127
Kolmogorov, 80

Lapicque, 8
Lyapunov, 130

Minkowski, 102

Nagumo, vii, 32
Newton, 4

Parkinson, 12
Poisson, vii

Quaglino, 129

Saviozzi, 127

Tirozzi, 119

Volt, 3
Voronoi, 115

Wiener, vii

Series on Advances in Mathematics for Applied Sciences

Series on Advances in Mathematics for Applied Sciences

Aims and Scope

This Series reports on new developments in mathematical research relating to methods, qualitative and numerical analysis, mathematical modeling in the applied and the technological sciences. Contributions related to constitutive theories, fluid dynamics, kinetic and transport theories, solid mechanics, system theory and mathematical methods for the applications are welcomed.

This Series includes books, lecture notes, proceedings, collections of research papers. Monograph collections on specialized topics of current interest are particularly encouraged. Both the proceedings and monograph collections will generally be edited by a Guest editor.

High quality, novelty of the content and potential for the applications to modern problems in applied science will be the guidelines for the selection of the content of this series.

Instructions for Authors

Submission of proposals should be addressed to the editors-in-charge or to any member of the editorial board. In the latter, the authors should also notify the proposal to one of the editors-in-charge. Acceptance of books and lecture notes will generally be based on the description of the general content and scope of the book or lecture notes as well as on sample of the parts judged to be more significantly by the authors.

Acceptance of proceedings will be based on relevance of the topics and of the lecturers contributing to the volume.

Acceptance of monograph collections will be based on relevance of the subject and of the authors contributing to the volume.

Authors are urged, in order to avoid re-typing, not to begin the final preparation of the text until they received the publisher's guidelines. They will receive from World Scientific the instructions for preparing camera-ready manuscript.

SERIES ON ADVANCES IN MATHEMATICS FOR APPLIED SCIENCES

SERIES ON ADVANCES IN MATHEMATICS FOR APPLIED SCIENCES

SERIES ON ADVANCES IN MATHEMATICS FOR APPLIED SCIENCES

SERIES ON ADVANCES IN MATHEMATICS FOR APPLIED SCIENCES